MW01556822

STUDY GUIDE

TO ACCOMPANY

INTRODUCTION TO

GENERAL, ORGANIC & BIOCHEMISTRY

sixth edition

Bettelheim • Brown • March

WILLIAM M. SCOVELL

Bowling Green State University

BROOKS/COLE

THOMSON LEARNING™

Australia • Canada • Mexico • Singapore • Spain • United Kingdom • United States

COPYRIGHT © 2001 Thomson Learning, Inc.
Thomson Learning™ is a trademark used herein under license.

ALL RIGHTS RESERVED. No part of this work covered by the copyright hereon may be reproduced or used in any form or by any means—graphic, electronic, or mechanical, including, but not limited to, photocopying, recording, taping, Web distribution, information networks, or information storage and retrieval systems—without the written permission of the publisher.

Printed in the United States of America

3 4 5 6 7   05 04 03 02 01

0-03-029234-4

> **For more information about our products, contact us at:**
> **Thomson Learning Academic Resource Center**
> **1-800-423-0563**
>
> **For permission to use material from this text, contact us by:**
> Phone: 1-800-730-2214
> Fax: 1-800-731-2215
> Web: www.thomsonrights.com

Asia
Thomson Learning
60 Albert Complex, #15-01
Alpert Complex
Singapore 189969

Australia
Nelson Thomson Learning
102 Dodds Street
South Street
South Melbourne, Victoria 3205
Australia

Canada
Nelson Thomson Learning
1120 Birchmount Road
Toronto, Ontario M1K 5G4
Canada

Europe/Middle East/South Africa
Thomson Learning
Berkshire House
168-173 High Holborn
London WC1 V7AA
United Kingdom

Latin America
Thomson Learning
Seneca, 53
Colonia Polanco
11560 Mexico D.F.
Mexico

Spain
Paraninfo Thomson Learning
Calle/Magallanes, 25
28015 Madrid, Spain

TO THE STUDENT

> I hear, and I forget
> I see, and I remember
> I do, and I understand
>
> "Ancient Chinese Proverb"

The aim of this Study Guide is simply to aid the student in increasing understanding of the material presented in Introduction to General, Organic and Biochemistry by Bettelheim and March, and to make studying more efficient. The general philosophy or approach put forth here should not only be useful in this course, but in many courses in which the learning of specific new concepts and the development of problem solving abilities is important. The initial focus of attention should be to determine the immediate objectives or aims of each chapter. Therefore study objectives will begin every chapter in order to point out aspects worthy of special emphasis in your studies.

Once the major objectives have been read over carefully, one must ask, "How best can I study to develop a firm understanding and a command of the new vocabulary, concepts and problem solving techniques associated with the material?" The approach to this end will be to define four study practices.

1. A list of key terms follows the summary at the end of each chapter in the text. Additional terms of importance from the text will be listed in the Study Guide either individually, or in cases in which two or more terms are related, they will be listed together. This latter grouping will serve to emphasize that the similarities and differences between these terms should be clearly understood. The terms are basic to the understanding of the new material and will provide the very foundation of your understanding. Review them together with the objectives before reading the chapter. After reading the text, reread the objectives and define each term. To deepen your understanding of the terms, include a specific example in each case where this is appropriate. Refer to the Glossary in the text to provide clear definitions to all the terms. **Continually review** the definitions before each study session or until the responses are clear and spontaneous.

2. In each chapter, a few concepts provide the basis for grasping many of the sometimes apparently diverse, but interrelated subjects presented. In some chapters, a selected number of concepts will be elaborated on in order to "pull together" some of the ideas and to deepen understanding of the material.

3. An integral part of chemistry involves the solving of both mathematical and conceptual problems. The general thought processes and interrelationships helpful in reaching these solutions will be briefly outlined. These problems will, in some cases, include the detailed solutions of some of the problems in the text.

4. At the point which the student feels confident that the material has been mastered, a **SELF-TEST** will provide a means for evaluating the progress toward reaching the major objectives in the chapter. Answers to all the questions and problems will be included, with explanations accompanying some of the answers.

NOTE: Answers to even-numbered, end-of-chapter problems in the text can be found on the Web site for Bettelheim, Brown & March: *General, Organic & Biochemistry, sixth edition*, under www.brookscole.com/chemistry_d

SOME OBSERVATIONS AND <u>A SUGGESTED STUDY APPROACH</u>

1. Typically students who obtain above average grades will read the chapter a minimum of twice, if not three times. I point it out only because I find many freshman are surprised to learn that this effort is actually the **norm** and to be sure, not unusual.

2. Reading the chapter prior to the material being discussed in class usually affords the student the unique advantage of being able to clarify questions on the material at the time the professor is introducing it to the class. Unfortunately, for many students, this often becomes very difficult, if not impossible. However, try it. You may find it so valuable that it may become a part of your personal study strategies.

3. Take good notes in class and as soon after class as possible, **recopy the notes into another notebook in a more organized fashion**. Do this at the same time that you refer to your textbook and annotate your class notes with extra definitions or comments from the text. I have found that this has been especially helpful to many students.

4. Knowledge of a subject begins with **memorizing** the meaning of new terms, concepts and equations. However, **understanding** the material goes far beyond the memorization of these elements in a "vacuum." It requires that the interrelationships between all these elements be clearly resolved. This requires a continual review of the material and a major input of time and effort into solving problems and answering focused questions. Therefore, it is important that you commit yourself to working as many problems as possible. There is no substitute for this.

I would like to outline a study procedure that many students have found helpful. This, of course, will not be for everyone. Each student must develop routines which serve him or her best. Try this approach if you wish. If it helps, use it.

1. **Review** the chapter objectives and important new terms in the study guide. This will alert you to the key ideas and concepts and help to underline their importance before your reading.

2. **Read** the (text) chapter with these points in mind.

3. **Record** good notes in class and as soon after class as possible, rewrite the notes. Clarify all questionable points in the lecture notes **AT THIS TIME**, with the aid of your text. For those concepts which are not clear, **ask the professor immediately** after the next lecture or bring up this point at his or her next office hour.

4. **Write out** the definitions of all important new terms.

5. **Review** any worked problems in the chapter, with the aim of defining the general thought processes involved in solving the problem.

6. Now, the assigned questions and problems at the end of each chapter in the text should be worked and the material in the Study Guide should be studied. The guide will be most useful in cases where some additional explanation or worked examples are helpful. In some cases, simply rephrasing of text material may bring out the meaning more clearly.

7. **Reread** the chapter, the rewritten class notes, chapter objectives, important terms and equations and the focused review section.

8. Return to doing questions and problems that were initially unclear or difficult and then do additional ones.

9. **Do the SELF-TEST** in the study guide.

TABLE OF CONTENTS

Chapter	Title	Page
1	Matter, Energy, and Measurement	1
2	Atoms	19
3	Chemical Bonds	23
4	Chemical Reactions	29
5	Gases, Liquids, and Solids	39
6	Solutions and Colloids	49
7	Reaction Rates and Equilibrium	55
8	Acids and Bases	61
9	Nuclear Chemistry	67
10	Organic Chemistry	73
11	Alkanes and Cycloalkanes	77
12	Alkenes and Alkynes	83
13	Alcohols, Ethers, and Thiols	91
14	Benzene and its Derivatives	97
15	Chirality	102
16	Amines	108
17	Aldehydes and Ketones	113
18	Carboxylic Acids, Anhydrides, Esters, and Amides	119
19	Carbohydrates	126
20	Lipids	131
21	Proteins	136
22	Enzymes	142
23	Chemical Communications: Neurotransmitters and Hormones	148
24	Nucleotides, Nucleic acids, and Heredity	153
25	Gene Expression and Protein Synthesis	160
26	Bioenergetics: How the Body Converts Food to Energy	166
27	Specific Catabolic Pathways: Carbohydrate, Lipid, and Protein Metabolism	176
28	Biosynthetic Pathways	181
29	Nutrition and Digestion	185
30	Immunochemistry	190
31	Body Fluids	193

"It is only the first step
which takes the effort"
    - Madame Marie Vichy-Deffand

Matter, Energy and Measurements

CHAPTER OBJECTIVES

After studying Chapter 1 and the material in the Appendix on significant figures, and working the assigned exercises in the text and the study guide, you should be able to do the following.

1. Distinguish between a fact, a hypothesis and a theory. Indicate how they are associated with the scientific method.

2. Convert large and small numbers to an equivalent form in exponential notation. Carry out the reverse procedure.

3. Perform addition, subtraction, multiplication and division calculations using exponential notation. Evaluate the answers to insure that there are the proper number of significant figures.

4. List the word prefixes used in the metric system and the corresponding values.

5. Define temperature and indicate temperature scales in common use. Write the equations which relate temperature in (i) °C to °F; (ii) °F to °C (iii) °C to K; perform temperature conversions from one scale to another.

6. Using the factor-label method, perform conversions between the metric and English units, English to metric units, in addition to conversions within the metric system.

7. List the three states of matter and the characteristics which are unique to each state.

8. Distinguish between density and specific gravity.

9. Define energy and list six different forms of energy pointed out in this chapter.

10. State the Law of Conservation of Energy.

Chapter 1 Matter, Energy, and Measurement

11. Define heat and specific heat and indicate the units of each.

12. Using the rules for determining the number of significant figures, evaluate the number of significant figures in numbers listed throughout the chapters.

13. Define the important terms and comparisons in this chapter and give specific examples where appropriate.

IMPORTANT TERMS AND COMPARISONS

Matter and Energy
Fact, Hypothesis and Theory
Coefficient and Exponent
Kilo- and Milli-
Mega- and Micro
Centi-
Mass and Weight
Significant Figures
Scientific Method
Chemical and Physical Properties
Chemistry

Centigrade, Fahrenheit and Kelvin Scale
States of Matter
Density and Specific Gravity
English vs. Metric System of Units
Potential and Kinetic Energy
Law of Conservation of Energy
Temperature and Heat
Calorie and Joule
Hypothermia and Hyperthermia
Exponential Notation
Measured and Defined Numbers

FOCUSED REVIEW OF CONCEPTS

A. Interrelationships

  (i) °F = 9/5(°C) + 32

  (ii) °C = 5/9(°F-32)

  (iii) K = °C + 273

  (iv) Density = mass/volume;   D = M/V
  D is a **physical quantity** and has **units**

  (v) Specific Gravity (S.G.) =

  Density of substance of interest
  Density of water

  S.G. = D_A/D_{H_2O}

  S.G. is a unitless number

  (vi) Heat (cal) = Specific heat (cal/g-deg) x mass (g) x (T_2-T_1)

It should be noted that both equations (i) and (ii) define the relationship between the temperature in degrees Centigrade and in degrees Fahrenheit. Therefore, you should be able to rearrange equation (i) to arrive at equation (ii).

Chapter 1 Matter, Energy, and Measurement

Write eqn (i)

$$°F = 9/5 \, (°C) + 32$$

Subtract 32 from each side

$$°F - 32 = 9/5(°C) + 32 - 32$$

$$(°F - 32) = 9/5(°C)$$

Multiple each side by 5/9

$$5/9(°F - 32) = (5/9)(9/5)(°C)$$

This is eqn (ii)

$$5/9(°F - 32) = °C$$

Equation (iii) should be put to memory since many calculations to be performed in subsequent chapters (e.g., Chapter 5) will require that the temperature be expressed in degrees Kelvin (K).

Density is an intensive property of a substance and has units of mass/vol, which although usually expressed in g/mL, can be expressed in others units (i.e., lb/in^3, etc.). Recall also that **in**tensive properties are those which are **in**trinsic characteristics of the substance and therefore do not depend on the quantity of the substance.

Specific gravity is a ratio of the density of any substance to the density of water at 20°C. Therefore, specific gravity is simply a number, with no units.

B. Significant Figures
A scientist must evaluate the number of significant figures in every measurement and in every calculation using measured numbers. The number of significant figures is defined as the number of digits that are known with certainty.
The rules are the following.
1. All digits written down are significant, except in the case of zeros.
    a. Zeros that come before the first non zero digit are not significant.
    b. Zeros that come after the last non zero digit are significant if the number is a decimal. However, if the number is a whole number (no decimal point), they may or may not be significant.
2. In multiplying or dividing numbers, the **final answer** should have the same number of significant numbers as there are in the number with the fewest significant numbers.
3. In adding or subtracting a group of numbers and arriving at the number of significant figures in the answer, the number of significant numbers in the individual numbers does not matter. The answer must contain the same number of decimal places as the number in the listing with the fewest decimal places.
4. In rounding off numbers, if the first digit dropped is a 5, 6, 7, 8 or 9, raise the last digit kept to the next higher number. In all other cases, do not raise the last digit.
5. Counted or defined numbers, such as the number of sides on a square, are treated as if they have an infinite number of zeros following the decimal point.

Chapter 1 Matter, Energy, and Measurement

PROBLEM SOLVING METHODS

A. <u>Exponential Notation</u>

It is important to be able readily to express large and small numbers in exponential or scientific notation. This requires that the number be written as a **coefficient**, with a value between 1 and 10, **multiplied** by 10 raised to some power. For example,

$100 = 1 \times 10^2 = 1 \times (100)$ The solid line underlines the number of places the decimal point has been moved in each case.

$20,000 = 2 \times 10^4 = 2 \times (10,000)$

$5,200,000 = 5.2 \times 10^6 = 5.2 \times (1,000,000)$

Note in these examples that when the LARGE number (greater than 1) is changed to exponential notation form, the decimal point is moved to the LEFT by 2, 4 or 6 places, respectively. The positive exponent simply expresses the number of places that the decimal has been moved to the left.

$0.001 = 1 \times 10^{-3} = 1 \times (0.001)$

$0.000043 = 4.3 \times 10^{-5} = 4.3 \times (0.00001)$

$0.0200100 = 2.001 \times 10^{-2} = 2.001 \times (0.01)$

Similarly, in changing these SMALL numbers (less than one) to exponential notation, the decimal point is moved to the RIGHT by 3, 5 or 2, respectively. (Negative exponent)

a. MOVING THE DECIMAL POINT in Numbers Already in Exponential Notation.

One of the most common difficulties that students encounter occurs when doing multiple step calculations. Midway through the calculation, numbers such as the following may appear.

$16,743 \times 10^{14}$ 0.00713×10^6 0.0000431×10^{-6}

To change these numbers to proper form (i.e., the coefficient being between 1 and 10), the decimal point must be moved. Which way is always the question!

The rule to remember is:

IF THE **COEFFICIENT IS DECREASED** in magnitude, the **EXPONENT MUST BE INCREASED ACCORDINGLY**.

IF THE **COEFFICIENT IS INCREASED**, the **EXPONENT IS DECREASED.**

For example:

$16,743 \times 10^{14} \rightarrow 1.6743 \times 10^{18}$

coefficient decreased, decimal moved **4** places
exponent increased by 4

Chapter 1 Matter, Energy, and Measurement

$0.00713 \times 10^6 \rightarrow 7.13 \times 10^3$

coefficient increased, decimal moved **3** places
exponent decreased by 3

$0.0000431 \times 10^{-6} \rightarrow 4.31 \times 10^{-11}$

coefficient increased, decimal moved **5** places
exponent decreased by 5

Exercise I: Carry out these manipulations to develop skill in these conversions.
Number   Exponential Notation (with coefficient between 1 and 10)

1. 0.00102×10^1 _____
2. 632×10^{-14} _____
3. 0.00700×10^{43} _____
4. $13{,}642{,}000 \times 10^{-1}$ _____
5. 11×10^{23} _____
6. $0.0000006 \times 10^{-14}$ _____
7. 137×10^{-10} _____
8. 0.006×10^0 _____
9. $7{,}423{,}000{,}000 \times 10^{17}$ _____
10. 0.4361×10^{31} _____

In the frequently used manipulations with numbers in scientific notation outlined below, a TWO-STEP procedure can be used in each case.

b. ADDITION OR SUBTRACTION in Exponential Notation

Procedure: (i) change the numbers so all have the same exponent,
irrespective of the size of the coefficient
(ii) Add (or subtract) the coefficients; the exponent remains the same

Ex. a. 230×10^{-8} (i) 23×10^{-7}
6.10×10^{-5} $\rightarrow$ 610×10^{-7}
(ii)   ↓   ↓
$633 \times 10^{-7} = 6.33 \times 10^{-5}$

b. 340×10^6 (ii) 340×10^6
$-\ 220 \times 10^5$ $\rightarrow$ $-\ 22 \times 10^6$
(ii)   ↓   ↓
$318 \times 10^6 = 3.2 \times 10^8$

c. MULTIPLICATION in Exponential Notation

Procedure: (i) Multiple the coefficients directly
(ii) Algebraically **add** the exponents

Chapter 1 Matter, Energy, and Measurement   6

                (iii) Combine coefficients and exponent
                (iv) If necessary, change the coefficient to a value between 1 and 10 and make the corresponding change in the value of the exponent.

Ex. a. $(6.0 \times 10^{-14})(3.0 \times 10^{6})$ (i)   →   18
                                                    (ii)   → $\times 10^{-8}$
                                                    (iii) 18×10^{-8}
                                                    (iv) $\underline{1.8 \times 10^{-7}}$

       b. $(1 \times 10^{-14})(7 \times 10^{-2})$ (i)   →   7
                                                     (ii)   → $\times 10^{-16}$
                                                   (iii & iv) $\underline{7 \times 10^{-16}}$

d.   **DIVISION** in Exponential Numbers

Procedure: (i) Divide the coefficients directly
                    (ii) Algebraically subtract the exponents
                    (iii) Combine
                    (iv) If necessary, change coefficient to a value between 1 and 10 and make the corresponding change in the value of the exponent.

Ex. a. $\dfrac{3 \times 10^{-7}}{1 \times 10^{7}}$ (i)   →   3
                            (ii)   → $\times 10^{-7-7}$
                            (iii & iv) $\underline{3 \times 10^{-14}}$

      b. $\dfrac{3 \times 10^{7}}{1 \times 10^{-7}}$ (i)   →   3
                            (ii)   → $\times 10^{7-(-7)}$
                          (iii & iv) $\underline{3 \times 10^{14}}$

      c. $\dfrac{2.14 \times 10^{-3}}{8.60 \times 10^{+4}}$ (i)   →   0.249
                            (ii)   → $\times 10^{-3-4}$
                          (iii & iv) $0.249 \times 10^{-7} = \underline{2.49 \times 10^{-8}}$

      d. $\dfrac{5.91 \times 10^{-4}}{1.61 \times 10^{-12}}$ (i)   →   3.67
                            (ii)   → $\times 10^{-4+12}$
                          (iii & iv) $\underline{3.67 \times 10^{8}}$

B.   Unit Conversions

    (i)   All conversions within the metric system differ by some power of 10 as reflected in the prefixes. For example, centi = 10^{-2}; milli = 10^{-3}; micro = 10^{-6}; nano = 10^{-9}. These conversions are accomplished simply by moving the decimal point. Memorize these prefixes and those in Table 1.2 (text).

Chapter 1 Matter, Energy, and Measurement

The strategy in these problems is to multiply the starting physical quantity (a number with a unit) by a conversion factor which will both change the number and also its units.

AT EACH STEP IN THE CONVERSION, ONE UNIT MUST CANCEL OUT. AT NO POINT IN THE PROCESS ARE THERE MORE UNITS THAN YOU STARTED WITH. This conversion process is continued until the desired unit (or units) is (are) obtained. In each case, clearly understand what the initial units are and also what the units are you want to convert to.

Ex. 1   Change 2 ft to inches.   Start with **feet**; Convert to **inches**

$$\frac{2 \text{ ft}}{1} \cdot \frac{12 \text{ in}}{1 \text{ ft}} \rightarrow \frac{24 \text{ in}}{1} = 24 \text{ inches}$$

starting physical quantity   conversion factor

Note that in this one-step conversion, 2 units cancel (ft) and only one unit (in., the desired unit) remains. If the correct conversion factor had been used, but incorrectly inverted, **no units would have canceled**. For example:

$$\frac{2 \text{ ft}}{1} \cdot \frac{1 \text{ ft}}{12 \text{ in}} \rightarrow \frac{2 \text{ ft}^2}{12 \text{ in}}$$

Units did not cancel. This incorrect set-up produced more units (ft²/in) than in the starting quantity (ft). This is your signal that the conversion factor is inverted.

Ex. 2   Convert 3.2 milliliters to nanoliters.

Start with **milliliters**; want **nanoliters**. Depending on the conversion factors you know, this problem and many others, can be done in one or perhaps a few steps.

Route 1: $\frac{3.2 \text{ mL}}{1} \cdot \left(\frac{10^6 \text{ nL}}{1 \text{ mL}}\right) = 3.2 \times 10^6 \text{ nL}$

Route 2: $\frac{3.2 \text{ mL}}{1} \cdot \frac{10^3 \text{ μL}}{1 \text{ mL}} \cdot \left(\frac{10^3 \text{ nL}}{\text{μL}}\right) = 3.2 \times 10^6 \text{ nL}$

Although 1 nL = 10^{-9} L = 10^{-6} mL = 10^{-3} μL, you should note that the conversion factors are:

$\left(\frac{10^9 \text{ nL}}{\text{L}}\right), \left(\frac{10^6 \text{ nL}}{\text{mL}}\right), \left(\frac{10^3 \text{ nL}}{\text{μL}}\right)$

In each case, the exponents are **positive** to show the number of nanoliters contained in the larger volume unit.

Exercise II: Convert the quantities shown below into a metric equivalent. Since all the quantities are in the metric system, the conversion requires only moving the decimal to the left or the right.

Chapter 1 Matter, Energy, and Measurement

1. 1.32 L _____ mL _____ μL _____ nL

2. 6.32 mg _____ g _____ kg _____ mg

3. 0.039 cm _____ m _____ km _____ μm

4. 0.36 μs _____ ms _____ s _____ ns

(ii) General conversions that are not within the metric system require the same procedure. The conversion factors used, however, do not simply move the decimal point (i.e., the exponent), but change both the coefficient and the exponent in the starting number.

Ex. 1 How many nickels are in $363.00?
Start with **dollars**; convert to **nickels**.

$$\frac{363 \text{ dollars}}{1} \cdot \frac{20 \text{ nickels}}{1 \text{ dollar}} = (363)(20) \text{ nickels} = 7,260 \text{ nickels}$$

Ex. 2 The density of water is 1 g/cc. What is this equivalent to in terms of lb/in^3?

Solution:
This is a multistep conversion in which 2 units are to be converted.
g → lb.
cc. → in^3.

These problems are done by **completely converting one unit**, and then the other.

Recall 1 cc = 1 cm^3 1 lb = 454 g   1 in = 2.54 cm

$$\frac{1.0 \text{ g}}{cm^3} \cdot \frac{1.0 \text{ lb}}{454 \text{ g}} \cdot \frac{2.54 \text{ cm}}{1 \text{ in}} \cdot \frac{2.54 \text{ cm}}{1 \text{ in}} \cdot \frac{2.54 \text{ cm}}{1 \text{ in}} = \frac{(2.54)^3 \text{ lb}}{454 \text{ in}^3}$$

conversion   conversion of
from g to lb cm^3 → in^3

$$= 3.6 \times 10^{-2} \frac{lb}{in^3}$$

Ex. 3 Which of the following is the largest mass, and which is the smallest?

a) 3.5×10^{-8} kg
b) 27 g
c) 4×10^3 mg
d) 3.4×10^6 μg

Solution:
The main difficulty in determining which quantity is the largest and which is the smallest occurs because each mass is expressed in a different unit. The first task then is to convert all the masses to the same unit. It is not important **which** units are finally compared (kg. with kg., g with g or other units). The question here will be answered by comparing the masses in units of **μgs.**

Chapter 1 Matter, Energy, and Measurement   9

a. $\dfrac{3.5 \times 10^{-8} \text{ kg}}{1} \quad \dfrac{10^3 \text{ g}}{1 \text{ kg}} \quad \dfrac{10^6 \text{ μg}}{1 \text{ g}} \quad = \quad 3.5 \times 10^{-8+3+6} \text{ μg}$

$\hspace{10em} = \quad 3.5 \times 10^1 \text{ μg}$

b. $\dfrac{27 \text{ g}}{1} \quad \dfrac{10^6 \text{ μg}}{1 \text{ g}} \hspace{6em} = \quad 2.7 \times 10^7 \text{ μg}$

c. $\dfrac{4 \times 10^3 \text{ mg}}{1} \quad \dfrac{10^3 \text{ μg}}{1 \text{ mg}} \hspace{4em} = \quad 4 \times 10^6 \text{ μg}$

d. $3.4 \times 10^6 \text{ μg} \hspace{8em} = \quad 3.4 \times 10^6 \text{ μg}$

The comparison is now between all masses expressed in μg.
a. 3.5×10^1 μg
b. 2.7×10^7 μg or 27×10^6 μg
c. 4×10^6 μg
d. 3.4×10^6 μg

$\hspace{4em} 3.5 \times 10^{+1} \text{ μg} \;<\; 3.4 \times 10^6 \text{ μg} \;<\; 4 \times 10^6 \text{ μg} \;<\; 2.7 \times 10^7 \text{ μg}$

$\hspace{10em}$ increasing mass $\quad \rightarrow$

Therefore,
  a. has the smallest mass (35 μg)

  b. has the largest mass (2.7×10^7 μg or 27,000,000 μg)
  Work this problem again for practice and convert all the masses to a different common unit.

Exercise III: Perform the following conversions from the English to the metric system of units or vice versa. These conversions require the use of conversion factors and the factor label method. Refer to Table 1.3 (text) for the less widely used conversion factors.

1. 1.0 ft   __12__ in   __30__ cm   _____ mile

2. 300 g   __10__ oz(avdp)   _____ lb   _____ μg

3. 6.3 fluid oz. _____ μL   _____ L   _____ qt

Ex.4   If the density of air is 1.25×10^{-3} g/cc, what is the mass of the air in a room that is 5.0 meters long, 4.0 meters wide and 2.2 meters high?

Solution:
  One approach to this problem is the following:

  (i)   The density is known in g/cc.
  (ii) Determine the volume of the room in meters cubed (m^3) and then convert volume to cc

Chapter 1 Matter, Energy, and Measurement 10

(iii) Knowing the density now in g/cc and the volume in cc, calculate the mass of the air in the room in grams.

(i) D = 1.25 x 10^{-3} g/cc
(ii) Determine the volume
Volume = length x width x depth = (5.0 m) (4.0 m) (2.2 m) = 44 m^3
Conversion of 44 m^3 to cc

$$\frac{44 \text{ m}^3}{1} \cdot \frac{10^2 \text{ cm}}{1 \text{ m}} \cdot \frac{10^2 \text{ cm}}{1 \text{ m}} \cdot \frac{10^2 \text{ cm}}{1 \text{ m}} = 44 \times 10^6 \text{ cm}^3 \text{ or } 4.4 \times 10^7 \text{ cc}$$

(iii) Determine the mass of air.

$$D = \frac{M}{V} \qquad M = (D)(V)$$
$$= (1.25 \times 10^{-3} \text{ g/cc})(4.4 \times 10^7 \text{ cc})$$

M = 5.5 x 10^4 g

THE CONCEPT OF SPECIFIC HEAT

The specific heat of a substance is an intensive property of a substance - that is, it does not depend on the quantity of the substance. It is a measure of how much energy (heat) is required to increase the temperature of one gram of a substance by 1°C.

Ex. 1   Consider the specific heat of 3 substances.

specific heat (cal/g-deg)	
water	1.00
iron	0.11
lead	0.038

a. Which substance requires the most heat to increase 1 g of the substance 1°C?

b. If equivalent amounts of heat (for ex., 100 cal) are added to 1 g of each substance, which substance will experience the largest increase in temperature? Which will experience the least increase?

c. If 60 cals. are added to 6 g of iron at 20°C, what will the final temperature be?

Solution:
a. Using the definition of specific heat directly, the values indicate that the specific heat for water requires 1.00 cal to increase 1 g of water by 1°C. The specific heat value for iron is about one tenth the value for water, while the value for lead is smaller yet. Therefore, by definition and simple comparison, water requires the most heat.

b. An equivalent amount of heat added to 1 g of each substance would increase the temperature of lead the most, while the increase in water would be the least. This is exactly the opposite order as observed in part a.

c. **Heat** absorbed by a substance will produce an increase in **temperature**. The magnitude of the increase will depend on the **specific heat** (S.H.) of the

Chapter 1 Matter, Energy, and Measurement

substance and the **mass**. As shown in the section 1.9 in the text, the equation relating these quantities is:

$$\text{heat} = (S.H.) \times (m) \times (T_2 - T_1)$$

This equation has 5 quantities (heat, S.H., m, T_2, T_1). In this problem and in others, you must first determine (i) which quantities are given and (ii) which quantity is to be determined (i.e., the unknown). Since only one quantity in this problem is to be determined, 4 quantities must be known.

Known quantities:
- heat = 60 cal
- m = 6 g
- S.H. (iron) = 0.11 cal/g-deg
- T_1 = 20°C
- T_2 = to be determined

Inserting these quantities into the equation

$$60 \text{ cal} = (0.11 \text{ cal/g-deg})(6g)(T_2 - 20°)$$

Rearrange this equation by dividing both sides by (0.11 cal/g-deg) (6 g) and then canceling on the right side to yield

$$\frac{60 \text{ cal}}{(0.11 \text{ cal/g-deg})(6 \text{ g})} = (T_2 - 20°C)$$

$$90° = T_2 - 20$$

Finally, add 20 to both sides to arrive at the final temperature, T_2.

$$110°C = T_2$$

Ex. 2 If 425 cal of heat are added to a sample of 52 g of copper at 25°C, what will be the final temperature?

$$\text{heat} = (S.H.) \times (m) \times (T_2 - T_1)$$

Given:
- heat = ~~1025 cal~~ 425
- m = 52 g
- S.H. (copper) = 0.092 (Table 1.3, text)
- T_1 = 25°C
- T_2 = to be determined

$$425 \text{ cal} = 0.092 \text{ (cal/g-deg)} \times (52 \text{ g}) \times (T_2 - 25°C)$$

$$\frac{425 \text{ cal}}{(0.092 \text{ cal/g-deg})(52 \text{ g})} = T_2 - 25°C$$

$$89° = T_2 - 25°C$$

$$114° = T_2$$

Chapter 1 Matter, Energy, and Measurement   12

SELF-TEST QUESTIONS

MULTIPLE CHOICE. In the following questions, select the correct answer from the choices listed. In some cases two or more answers will be correct.

1. Indicate which of the following are extensive properties of a substance - that is, a property that depends on the quantity of the substance.
   a. mass
   b. density
   c. length
   d. specific heat
   e. kinetic energy
   f. melting point

2. Indicate whether each statement is a fact, hypothesis or a theory.
   a. Smoking cigarettes is detrimental to your health.
   b. The ocean water at Miami, Florida contains more salt than the water in Galvaston, Texas.
   c. A micrometer (um) is larger than a nanometer (nm).
   d. Judy received a higher grade on the exam because she studied longer.

3. Which of the following are forms of matter?
   a. thought
   b. nucleus of an atom
   c. a vacuum
   d. a line drawn on a paper
   e. a shadow

4. The number 2.2×10^{-4} is equivalent to:
   a. 0.0022
   b. 22×10^{-5}
   c. 22×10^{-3}
   d. 0.022×10^{-2}

5. The number 1×10^{-4} is equivalent to:
   a. $\dfrac{1}{1 \times 10^4}$
   b. 10×10^{-3}
   c. 0.0001
   d. 100×10^{-6}

6. Which of the following numbers can be added directly in exponential notation?
   a. 3.26×10^{-4}
   b. 600×10^{-2}
   c. 32.6×10^{-3}
   d. 0.40×10^{-4}

7. Multiplication of $(3 \times 10^{-7})(1 \times 10^6)(2 \times 10^0)$ equals:
   a. 3×10^{-1}
   b. 6×10^{-1}
   c. 6
   d. 60×10^{-3}

8. The answer to the problem $\dfrac{93 \times 10^4}{3.0 \times 10^{-4}}$ is:
   a. $31 \times 10^{+8}$
   b. 31
   c. 13
   d. 31×10^0

9. The answer to the problem $\dfrac{(3 \times 10^{-7})(5 \times 10^2)}{3 \times 10^1 + 20}$ is:
   a. 5×10^{-4}
   c. 3×10^{-4}

Chapter 1 Matter, Energy, and Measurement

    b. 3×10^{-6} d.   0.005

10. Which of the following are equivalent to 50 mL?
    a.   0.5L   c.   0.05 L
    b.   50 cc   d. 5×10^4 μL

11. Which of the following are equivalent to .5 ms?
    a.   500 μs   c. 8.3×10^{-6} min
    b. 5×10^{-4}s   d.   5000 μs

12. A degree on the Centigrade scale is:
    a.   larger than the degree on the Fahrenheit scale
    b.   is the same size as the degree on the Kelvin scale
    c.   9/5 larger than the Fahrenheit scale
    d.   none of the above

13. The temperature of -40°C corresponds to:
    a.   313 K   c.   -40°F
    b.   233 K   d.   -73°F

14. The temperature of 0 °K corresponds to:
    a.   273°C   c.   -459.4°F
    b.   absolute zero   d.   32°F

15. A density of 1.5 g/cc is equivalent to:
    a.   1500 g/L   c. 3.3×10^{-3} lb/mL
    b.   13.4 lb/ft^3 d.   6

16. How much energy is required to warm 15.0 g of ethyl alcohol from 4°C to 25°C? The specific heat of ethyl alcohol is 0.590 cal/g-deg.
    a.   158 cal   c.   0.185 kcal
    b.   185 cal   d.   37 cal

17. Which of the following ratios may be used directly as a conversion factor in converting microliters (μL) to liters (L)?
    a. $\dfrac{1 \text{ gal}}{3.785 \text{ L}}$ b. $\dfrac{1 \text{ L}}{10^{+6} \text{ uL}}$ c. $\dfrac{10^6 \text{ μL}}{1 \text{ L}}$ d. $\dfrac{10^{-6} \text{ uL}}{1 \text{ L}}$

18. Indicate the number of significant figures in the following measured numbers.
    a.   0.02   b.   32.00032   c.   0.00011   d.   120,000

COMPLETION. Write the word, phrase or number in the blank that will complete the statement or answer the question.

1. Matter is defined as _____.

2. Scientists proposed new _____ to be tested and continually examine the validity of _____ in more depth.

Chapter 1 Matter, Energy, and Measurement

3. In the number, 4×10^{-4}, the number 4 is called the _____, while -4 is the _____.

4. Another way of writing the following numbers is:
   a. $\dfrac{1}{10^{-4}}$ = _____   c. 43×10^6 = _____
   b. 0.001 = _____   d. 9,430,000,000 = _____

5. The number $87,000 \times 10^{-16}$ is the same as 8.7×10^{-n} in which n equals _____.

6. SI is the abbreviation for _____. This system is based on the _____.

7. In the SI, the basic unit of length is the _____, which is abbreviated as _____; the basic unit of volume is _____, which is abbreviated as _____.

8. A kilometer is _____ times longer than a centimeter.

9. Scientists use a balance to experimentally determine the _____ of a substance.

10. The basic unit for _____ is the same in the English and metric systems. This abbreviation is _____.

11. The size of the degree on the Kelvin scale is _____ to that on the Centigrade scale.

12. A temperature of 300°C corresponds to _____ K.

13. A temperature of 37°C is equal to _____ °F.

14. Absolute zero is defined as _____ K. At this temperature, the motion of molecules _____.

15. There are _____ L in 6.7 gallons of wine.

16. The three states of matter are _____, _____ and _____.

17. To change a liquid to a gas, _____ must be added to the liquid.

18. To change a liquid to a solid, energy must be _____ from the liquid.

19. For the three states of matter, the order of compressibility is _____ > _____ > _____.

20. Oil spilled by large tankers floats on the ocean because oil is _____ than ocean water.

21. The density and the specific gravity of a substance differ in that _____.

22. Density and specific gravity are _____ properties of a substance.

23. The density of a substance _____ as the temperature is increased.

24. A hydrometer is used to measure the _____ of a liquid.

25. Energy is defined as _____.

Chapter 1 Matter, Energy, and Measurement

26. Kinetic energy is the energy of _____, while potential energy is _____ energy.

27. The law of conservation of energy states that _____.

28. A high value for the specific heat of a substance indicates that _____.

29. Substance 1 has a specific heat of 3.0 cal/g-deg, while substance 2 has a specific heat of 1.0 cal/g-deg at 20°C. The addition of an equivalent amount of heat to 10 grams of either substance will result in substance _____ increasing more in temperature.

30. In changing a large number into exponential notation form, the decimal point is moved to the _____ by n places and the exponent is equal to _____.

31. A temperature of 150 K is equal to _____ on the Fahrenheit scale.

32. A volume of 131 mL of water weighs _____.

33. A 15.0 mL volume of liquid with a specific gravity of 1.75 weighs _____.

34. A volume of a clear blue liquid with a density of 0.89 g/cc weighs 43 g. The volume is _____. Its specific gravity is equal to _____.

35. A copper atom weighs 2.3×10^{-25} lb. Its weight in milligrams is _____, and its corresponding weight in kilograms is _____.

36. A drop or rain has a volume of about 50 microliters. This is equivalent to _____ quarts (1 quart = 0.946 L).

37. A single bond between two carbon atoms has a length of 1.54 Angstroms (1 Angstrom = 1×10^{-10} m). This bond length is equivalent to _____ in.

38. A substance is found to have a density of 2.40 g/cc and weighs 100. g. Its volume is _____.

39. At 20°C, the specific gravity of a solid substance A is 2.46, while the density of a liquid B is 2.0 g/cc. If solid A is placed in liquid B, it will _____.

40. A block of Blasa wood has dimensions of 2.00 cm x 6.00 cm x 1.50 m and a density of 0.120 g/cc. Its mass is _____.

41. Express the following quantities using the metric (SI) units indicated.
    a. 34 cm = _____ m
    b. 1×10^3 g = _____ µg
    c. 2.3×10^3 g/cc = _____ mg/L
    d. 6.3 cal/g-deg = _____ kcal/mg-deg

42. Volumes of liquids may be measured accurately with either a _____, _____ or _____. When only approximate volumes are needed, a _____ is used.

43. The _____ of an object is independent of location, while the _____ of an object is dependent on location.

Chapter 1 Matter, Energy, and Measurement   16

44. The _____ of a substance is measured on a laboratory balance.

45. The density of a substance usually _____ with increasing temperature because the _____ increases with temperature, while the _____ does not change with temperature.

46. A billion seconds is equivalent to ca. ____ years. A billion seconds ago, most students in this class were probably _____ born yet.

47. Indicate the number of significant figures in the following
    a. 2.00   _____
    b. 11,000   _____
    c. 0.0043   _____
    d. 5.32×10^4 _____

    e. 5.0×10^{-3} _____

48. Carry out the following operations and report the correct number of significant figures in the answer.
    a. 14.00   b. 3,640.64
         1.042   - 13.1
         307.

    c. $0.37 \times 41.56 =$ _____

    d. $\dfrac{18.4}{3.468} =$ _____

    e. $\dfrac{0.1340 \times 6.20}{146.00} =$ _____

49. When heat is added to a substance, the temperature of the substance increases, with the magnitude of the temperature rise depending on the specific heat of the substance. Distinguish between the terms temperature, heat and specific heat by writing out the definition for each.

ANSWERS FOR FOCUSED REVIEW EXERCISES

Exercise I
1. 1.02×10^{-2}
2. 6.32×10^{-12}
3. 7.00×10^{40}
4. 1.3642×10^6
5. 1.1×10^{24}
6. 6×10^{-21}
7. 1.37×10^{-8}
8. 6×10^{-3}
9. 7.423×10^{26}
10. 4.361×10^{30}

Exercise II
1. 1.32×10^3 mL; 1.32×10^6 µL; 1.32×10^9 nL
2. 6.32×10^{-3} g; 6.32×10^{-6} kg; 6.32×10^{-9} Mg
3. 3.9×10^{-4} m; 3.9×10^{-7} km; 3.9×10^2 µm

Chapter 1 Matter, Energy, and Measurement 17

4. 3.6×10^{-4} ms; 3.6×10^{-7} s; 3.6×10^{2} ns

Exercise III
1. 12 in; 30 cm; 1.9×10^{-4} miles
2. 10 oz; 0.7 lb; 3×10^{8} µg
3. 1.9×10^{5} µL; 1.9×10^{-1} L; 2.0×10^{-1} qts

ANSWERS TO SELF-TEST QUESTIONS

Multiple Choice
1. a, c, e
2. a (fact); b & d(hypothesis); c(fact)
3. b, d
4. b, d
5. a, c, d
6. a, d
7. b
8. a
9. b
10. b, c, d
11. a, b, c
12. a, b, c
13. b, c
14. b, c
15. a, c
16. b, c
17. b; in this conversion, L must be in the numerator
18. a. one   b. seven   c. two
    d. ambiguous. Can be two or as many as six.

Completion
1. anything that has mass and also occupies space
2. hypotheses, theories
3. coefficient, exponent
4. a. 1×10^{4}
   b. 1×10^{-3}
   c. 4.3×10^{7}
   d. 9.43×10^{9}
5. 12
6. System International, Metric System
7. meter, m, liter, L
8. 10^{5}
9. mass
10. time, s
11. identical
12. 573 K
13. 98.6°F
14. 0 K or -273°C, ceases
15. 25 liters
16. solid, liquid, gas
17. heat energy
18. taken away
19. gas > liquid > solid
20. less dense
21. density is a physical quantity and has units, while specific gravity is a ratio of densities and is unitless
22. **in**tensive
23. decreases

24. specific gravity
25. the capacity to do work
26. motion; stored
27. energy can neither be created or destroyed
28. the input of large amounts of energy or heat is necessary to increase the temperature of the substance
29. substance 2
30. left, plus n
31. $-253°F$
32. 131 g
33. 26.2 g
34. 48 cc (mL), 0.89
35. 1.0×10^{-19} mg, 1.0×10^{-25} kg
36. 5.3×10^{-5} qt
37. 6.06×10^{-9} in
38. 42 cc (mL)
39. sink
40. 216 g
41. a. 0.34 m

    b. $1 \times 10^{+9}$ μg

    c. 2.3×10^9 mg/L

    d. 6.3×10^{-6} cal/mg-deg
42. buret, pipet, volumetric flask, graduated cylinder
43. mass, weight
44. mass
45. decreases, volume, mass
46. 31.7 years. In 1997, a billion seconds ago would be about the year 1965-66. Most students in this class were not born yet.
47. a. 3
    b. at least 2; if all figures are significant, then 5
    c. 2
    d. 3
    e. 2
48. a. 322
    b. 3627.5
    c. 15
    d. 5.31
    e. 0.00569
49. Temperature is the hotness or coldness of a substance, measured by one of the temperature scales. Heat is a form of energy that is measured in calories, with a calorie being the amount of heat that is necessary to raise 1 g of water by $1°$ C (centigrade). The specific heat is an inherent property of every substance. It is the amount of heat, in calories necessary to raise the temperature of 1 g of the substance by $1\ °C$.

Learning in old age is writing on sand but
learning in youth is engraving on stone.
 - Arabian proverb

Atoms

CHAPTER OBJECTIVES

This chapter begins to present the fundamental aspects of chemistry. Whether or not you have been introduced to most of the terms and concepts previously in high school, you should very carefully master the material since it provides the real starting point and the basic grounding for subsequent chapters. You will find that an understanding of the biochemistry associated with a living organism relies heavily on the basic concepts and principles of chemistry.

After studying the chapter and working the assigned exercises in the text and the study guide, you should be able to do the following.

1. Summarize and contrast the ideas of Zeno, Democritus, Dalton and Bohr regarding the basic makeup of matter.

2. Name and write out the chemical symbols for the first twenty elements (this is a minimum objective which should be extended).

3. Distinguish between a pure compound and a mixture.

4. State the four basic principles which serve as a basis for Dalton's atomic theory.

5. Characterize the elementary particles in the atom, including their charge, mass and location within the atom.

6. Given the number of protons and neutrons in an atom, determine the charge of the nucleus and the atomic number and mass number for the element.

7. Given the atomic number of an element, locate and identify the element on the periodic table.

8. Distinguish between an atom, isotope, a positive and negative ion and a molecule.

9. Explain the difference between a representative element and a transition metal.

10. Write out the ground state electronic configuration for at least the first 40 elements, making

Chapter 2 Atoms

use of the periodic table, the Pauli Exclusion Principle and Hund's Rule.

11. Write out the ground state electronic configuration for positively and negatively charged ions of the first 20 elements.

12. Explain the trend in the ionization energies of the elements in a column (i.e., a group) and in a horizontal row on the periodic table.

13. List the four most abundant elements in (a) the earth's crust and those in (b) the human body.

14. Define the important terms and comparisons in this chapter and give specific examples where appropriate.

IMPORTANT TERMS AND COMPARISONS

Atom, Ion and Molecule
Element
Compound and Mixture
Representative Element and Transition Metal
Periodic Table
Dalton's Atomic Theory
Law of Conservation of Mass
Ground State and Excited State
Law of Constant Composition
Shell and Subshell
Unpaired and Paired Electrons
Inner and Outer Shell
Electronic Configuration
Monatomic, Diatomic and Polyatomic Elements
Pauli Exclusion Principle

Atomic Weight
Halogen
Group Numbers and Group Names
Metal, Metalloid and Nonmetal
Isotopes
Noble Gases
Quantized Energy Levels
Ionization Energy
s, p, and d Orbitals
Proton, Neutron and Electron
AMU
Nucleons and Nucleus
Mass Number and Atomic Number
Hund's Rule
Lewis Structure

SELF-TEST QUESTIONS

MULTIPLE CHOICE. In the following exercises, select the correct answer from the choices listed. In some cases, two or more answers will be correct.

1. The concept of quantized energy levels in the atom was first proposed by:
    a. Zeno
    b. Bohr
    c. Lavoisier
    d. Democritus

2. Dalton's atomic theory includes the following statements:
    a. All atoms of the same element are identical to each other.
    b. All atoms are colorless.
    c. Atoms combine to form molecules. The molecule is a tightly bound group of atoms that acts as a unit.
    d. All matter is made up of very tiny indivisible particles called atoms.

3. The mass number of an atom which has 20 neutrons and a charge of +18 in the nucleus is:
    a. 20
    b. 38
    c. 56
    d. cannot determine

4. The atomic number of the atom in question 3 is:
    a. 20
    b. 38
    c. 18
    d. cannot determine

Chapter 2 Atoms

5. The ground state electronic configuration for the atom which has 6 protons and 6 neutrons in the nucleus is:

	1s	2s	2p$_x$	2p$_y$	2p$_z$	3s
a.	↓↑	↓↑	↓↑	☐	☐	☐
b.	↓↑	↓↑	↓	↓	☐	☐
c.	↓↑	↓↑	↓↑	↓↑	↓↑	↓↑
d.	↓↑	↓↑	↓	☐	☐	↓

6. A possible excited state electronic configuration for the atom in question 5 is answer _____.

7. Which of the following elements is a metalloid?
   a. Hydrogen, H
   b. Germanium, Ge
   c. Silver, Ag
   d. Helium, He

8. Which of the following collection of elements are nonmetals?
   a. Na, K, Rb
   b. F, Cl, Br
   c. O, S, Se
   d. Cu, Ag, Au

9. The four most abundant elements in the earth's crust, listed in increasing order, are:
   a. Ca < Fe < O < Si
   b. Fe < Al < Si < O
   c. N > C > H > O
   d. Cl < O < N < P

10. The mass of a single atom of lead (Pb) is:
    a. 3.5 g
    b. 3.5 ng
    c. 3.5×10^{-22} g
    d. 3.5×10^6 g

11. The four most abundant elements in the human body are:
    a. H, Li, C & N
    b. C, N, O & H
    c. C, N, S, & H
    d. Fe, C, N & Na

12. The four most common elements in the human body are considered:
    a. transition metals
    b. metalloids
    c. representative elements
    d. metals

13. The Lewis structure for sulfur is:
    a. :S̈: (6 dots)   b. :S   c. :S: d. :S̈: (8 dots)

COMPLETION. Write the word, phase or number in the blank which will complete the statement or answer the question.
1. The three elementary particles in an atom are the _____, _____ and _____, with only _____ of them being classified as nucleons.

Chapter 2 Atoms

2. Of the alkali metals, Li, Na, K, Rb and Cs, _____ requires the least energy to ionize an electron off and to produce a positive ion.

3. The chemical symbols for the elements beryllium, arsenic, potassium and sodium are _____, _____, _____ and _____, respectively.

4. The atomic nucleus always possesses a _____ charge because the _____ resides in the nucleus.

5. Isotopes of an element have the same number of _____ , but different number of _____. Therefore, their _____ numbers are the same although their _____ numbers differ.

6. A positive ion is produced from an atom by _____ an _____ the atom.

7. The halogens are in Group _____ and have _____ electrons in their outer shell.

8. Calcium, with atomic number 20, is in Group 2A. It is an example of a _____ element which has _____ electrons in its outer shell.

9. The ionization energy of elements in Group 1A _____ going down the column, while the ionization energy going across a horizontal row of elements (from left to right) generally becomes _____.

10. The periodic table works because elements in the same _____ have the same _____ in the _____ shell.

11. In going down Group 7A from fluorine to astatine, the molecular weights progressively increase. Generally, as the molecular weights of similar compounds increase, the boiling points and melting points _____ .

ANSWERS TO SELF-TEST QUESTIONS

Multiple Choice
1. b
2. a, c, d
3. b
4. c
5. b
6. a or d
7. b
8. b, c
9. b
10. c
11. b
12. c
13. a

Completion
1. proton, neutron, electron, 2
2. Cesium (Cs)
3. Be, As, K and Na
4. positive, proton
5. protons and electrons, neutrons, atomic, mass
6. pulling off or ionizing, electron from
7. 7, 7
8. representative or a metallic, 2
9. decreases, greater
10. group or column, number of electrons, outer
11. increase

Chapter 3 Chemical Bonds

"I believe the chemical bond is not so
simple as some people seem to think."
     - Robert S. Mulliken

3

Chemical Bonds

CHAPTER OBJECTIVES

Although much remains to be learned about the atom, the titles of this and all subsequent chapters reveal that a great deal of attention is devoted to understanding the chemistry of ions and molecules. In either case, the outer **electronic configuration of the atom is altered** by the gain or loss of an electron to form a charged particle, or alternatively, electrons are shared in a common covalent bond between two atoms. As a result, the chemical and physical properties of these chemical species are different than the atoms from which they were derived. Also, in the case of complex ions or molecules, the species take on specific three-dimensional shapes, unlike that of the beautifully simple spherical shape of an atom. The molecular shapes can be predicted from the VSEPR model.

After you have studied this chapter and worked the exercises in the text and study guide, you should be able to do the following.

1. Write the equations for the formation of a cation or anion from the parent atom.

2. Using the octet rule as a guide, predict why certain ions are stable, while others are nonexistent.

3. From a knowledge of the charge on the positive and negative ions and using the principle of electrical neutrality, predict the formulas for a variety of ionic compounds.

4. Distinguish between an atom, an ion, and a molecule.

5. Describe the characteristic difference between an ionic bond and a covalent bond.

6. Characterize a single, double, and triple bond and give specific examples of compounds containing these bond types.

7. Write the molecular formula for a variety of compounds, in addition to drawing the structural formula and Lewis structure for each compound.

Chapter 3 Chemical Bonds

8. State the VSEPR Model. Given the number of valence-shell electron pairs of the central atom of a molecule, predict the shape of the molecule and the size of the angle between the bonds in each. Give specific examples of molecules with each shape.

9. Define electronegativity and explain the trend in electronegativities in the periodic table as shown in Table 3.2 (text).

10. Draw the structural formulas for some di-, tri- and tetraatomic molecules. Given the electronegativity value for each atom, indicate whether the individual bonds and whether the molecules themselves are polar or nonpolar.

11. Using the general rules which relate the electronegativity differences of the atoms in a bond to the bond type, predict whether the bonds in a variety of diatomic molecules are ionic, polar-covalent or nonpolar-covalent.

12. Write the name, together with structural formulas for some common compounds (Table 3.1 and 3.6, text) and some medicinally important polyatomic ions (Box 3A, text).

13. Write out the names and chemical formulas for some of the inorganic compounds in the chapter.

14. Define the important terms and comparisons in this chapter and give specific examples where appropriate.

IMPORTANT TERMS AND COMPARISONS

Bond Angle
Covalent Bond
Electrical Neutrality
Electronegativity
Hemoglobin
Ion, Cation, Anion
Ionic and Covalent Bonds
Linear and Angular Molecules
Molecular and Structural Formula
Partial Charges
Polar and Non-Polar Molecule
Shared and Unshared Electron Pair
Single, Double and Triple Bond
Three-Dimensional Molecular Structure
Trigonal, Pyramidal, and Tetrahedral Shapes
VSEPR Model

SELF-TEST QUESTIONS

MULTIPLE CHOICE. In the following exercises, select the correct answer from the choices listed. In some cases, two or more answers will be correct.

1. The formula for cesium phosphate is:
   a. $Cs_3(PO_4)$
   b. $Cs_2(P_4)$
   c. $Cs(PO_4)_3$
   d. $CsPO_4$

2. The Lewis structure for the ozone molecule, O_3, is:

Chapter 3 Chemical Bonds   25

a. [structure: O with two O bonded below]

c. Ö=O=Ö

b. :Ö—Ö=O:

d. [structure: O at top, double bond to two O's below]

3. Which of the following ions is stable?
   a. F+   d. O+2
   b. S2-   e. I-
   c. Ca+   f. N3-

4. A particular atom has the following characteristics. The atomic number is 12 and it forms cations with a +2 charge. This element is:
   a. Na   c. S
   b. Al   d. Mg

5. Of the following bonds, the one with the greatest dipole moment is:
   a. O-H   c. B-H
   b. S-H   d. F-H

6. The order of decreasing dipole moments in the bonds in question 6 is:
   a. H-F > O-H > S-H > B-H
   b. H-F > S-H > B-H > O-H
   c. O-H > H-F > S-H > B-H
   d. S-H > O-H > H-F > B-H

7. Which of the following molecules have a correct Lewis structure?

   a. [H₂C=C(H)—Cl: structure]   c. [:C≡N:]⁻¹

   b. [ClO₄⁻ structure]   d. :C≡O:

8. Indicate which of the following compounds are covalent.
   a. CaCl2   c. SO2
   b. ClBr   d. KCl

9. Indicate which pair of atoms has the greatest electronegativity difference.
   a. H & C   c. F & Cl
   b. H & O   d. C & N

10. Predict the molecular formula for the compound that contains calcium and phosphate.
    a. Ca(PO4)   c. Ca2(PO4)3

Chapter 3 Chemical Bonds   26

    b.   Ca3(PO4)2   d.   Ca(PO4)2

COMPLETION. Write the word, phrase or number in the blank that will complete the statement or answer the question.

1. Based on the VSEPR model, determine the molecular shape of each molecule.
   a. AlF_4^-
   b. C_2HCl
   c. H_3O^+
   d. PCl_3
   e. BI_3
   f. I_3^-
      (The central I does not obey the octet rule.)

2. Classify whether the bond between the two elements shown is non-polar, covalent, polar covalent or ionic.
   a. C, C
   b. Li, F
   c. Na, O
   d. H, I
   e. O, S

3. Name these compounds or ions.
   a. CN^-

   b. $(NH_4)_2SO_4$

   c. NO_2^-

   d. K_2O

   e. $MgCl_2$

   f. $Li(CH_3COO)$

4. Indicate the bond angles in these molecules.

   a. CO_2

   b. CCl_4

   c. H_2O

   d. BCl_3

   e. C_2H_2

5. Write the chemical formulas for these compounds.
   a. sodium bicarbonate
   b. rubidium sulfate
   c. magnesium permanganate

Chapter 3 Chemical Bonds

  d. ammonium oxide _____

6. The formation of a positively charged _____ from an atom requires the _____ of an electron, while the formation of a negatively charged _____ requires the _____ of an electron.

7. A _____ bond involves the _____ of one or more electron pairs, while an _____ bond results from an electrostatic attraction between _____ particles.

8. Although the C^{+4} cation has a complete outer shell, it does not exist because _____ are unstable.

9. In solid NaCl, each Na^+ has _____ Cl^- ions as _____ _____.

10. Name these compounds.
    a. $NiCl_2$ _____
    b. $Cr(NO_3)_3$ _____
    c. $NaH(SO_4)$ _____
    d. MgF_2 _____

10. Write the chemical formula for each of these compounds.
    a. copper(II) iodide _____
    b. sodium oxide _____
    c. aluminum oxide _____

ANSWERS TO SELF-TEST QUESTIONS

Multiple Choice
1. a
2. a (note that 18 valence electrons are involved and the octet rule must be obeyed)
3. b, e, f
4. 4d
5. d
6. a
7. b, c, d
8. b, c
9. b
10. c

Completion
1. a. tetrahedral
   b. linear
   c. pyramidal
   d. pyramidal
   e. trigonal planar
   f. linear
2. a. nonpolar covalent, $\Delta E = 0$
   b. ionic, $\Delta E = 3.0$
   c. ionic, $\Delta E = 2.4$
   d. nonpolar covalent, $\Delta E = 0.4$
   e. polar covalent, $\Delta E = 1.0$
3. a. cyanide anion (or ion)
   b. ammonium sulfate
   c. nitrite anion (or ion)
   d. potassium oxide
   e. magnesium chloride

Chapter 3 Chemical Bonds

      f. lithium acetate
4. a. 180°
    b. 109.5°
    c. 105°
    d. 120°
    e. 180°
5. a. $Na(HCO_3)$
    b. $Rb_2(SO_4)$
    c. $Mg(MnO_4)_2$
    d. $(NH_4)_2O$
6. cation, removal, anion, addition
7. covalent, sharing, ionic, charged
8. concentrated charges on one atom
9. 6, nearest neighbors
10. a. nickel(II) chloride
    b. chromium(III) nitrate
    c. sodium hydrogen sulfate
    d. magnesium fluoride
11. a. CuI_2
    b. Na_2O
    c. Al_2O_3

....chemical transformations....are
responsible for corrosion and decay,
for development, growth and life.
  - C.N. Hinshelwood
    The Kinetics of Chemical Change, 1940

4

Chemical Reactions

CHAPTER OBJECTIVES

Chemistry is a quantitative science. This means that it is possible to determine the exact quantity of a substance - for example, the volume, density, mass or number of molecules of water in a raindrop. Importantly, one can relate **macroscopic** quantities, such as the mass of a substance, to **microscopic** quantities, such as the number of atoms or molecules in a given quantity of the substance. In addition, exact relationships can be represented by a balanced chemical equation or a mathematical equation. For example, we all realize that it requires gasoline to run an automobile. This process of combustion of the gasoline can be described quantitatively by a balanced chemical equation. This chapter presents these concepts, relationships and procedures which provide the basis for some of the quantitative aspects of chemistry.

After you have studied the chapter and worked the exercises in the text and study guide, you should be able to do the following.

1. Calculate the formula weight of any ionic or molecular compound from its molecular formula.

2. Calculate the number of moles in a given mass of a compound.

3. Using Avogadro's number of 6.02×10^{23}, determine the number of molecules in a given mass of a compound.

4. Readily interconvert between the mass of a substance and the number of moles and molecules of the compound.

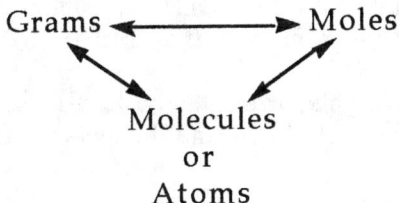

5. Using the atomic weights, determine the relative mass of pairs of atoms on the periodic table.

6. Balance simple chemical equations.

Chapter 4   Chemical Equations

7. From a balanced chemical equation, carry out calculations to determine the mass and molar relationship between reactants and products.

8. Using the balanced equation and given the actual quantity of reactants and of the product formed in a reaction, calculate the percentage yield of product.

9. Write out the complete, balanced equations for the reaction of various ionic compounds in aqueous solution. Point out the spectator ions and then rewrite the ionic equation.

10. Write out the general reaction scheme for a combination, decomposition, and a single- and double displacement reaction and then give a specific example of each.

11. Be able to recognize and balance a redox reaction. Indicate the species which are oxidized and reduced and the species which are the oxidizing and the reducing agent, respectively.

12. Explain how a voltaic cell works.

13. Given balanced equations including the heat gain or loss, generally classify the reaction type and indicate whether the reaction is exothermic or endothermic.

14. Define the important terms and comparisons in this chapter and give specific examples where appropriate.

IMPORTANT TERMS AND COMPARISONS

Chemical Reaction
Reactant and Product
Formula Weight and Molecular Weight
AMU
Avogadro's Number and Mole
Factor-Label Method
Conversion Factor
Balanced Equation
Coefficient and Subscript
Conservation of Mass
Actual Yield and Theoretical Yield
Side Reaction
Aqueous
Combination and Decomposition Reactions

Symbols, ↑, ↓
Oxidation and Reduction
Oxidizing and Reducing Agent
Combustion
Antiseptic and Disinfectant
Respiration
Rusting
Metabolism
Heat of Reaction and Heat of Combustion
Exothermic and Endothermic
Exergonic and Endergonic
Calorie
Displacement Reactions

FOCUSED REVIEW

To become comfortable and efficient with the mathematical manipulations of the new material, work the following exercises before going onto the SELF-TEST QUESTIONS.

Exercise I.   Calculation of Formula Weights. From the molecular formulas, calculate the formula weight of each compound. Round the answer off to the first decimal place.

a. $Mg(ClO_4)_2$ _____

b. Na_2O _____

Chapter 4   Chemical Equations

c. Al(H₃CCO₂)₃ _____

d. FeBr₃ _____

e. O₃ _____

f. C₁₈H₃₆O₂ _____

g. C₃H₇NO₂ _____

h. Pt(NH₃)₂Cl₂ _____

Exercise II.   Interconversion of Grams, Moles, and Molecules. The most important interrelationship in this chapter is that between grams, moles and molecules and is represented below.

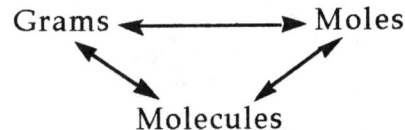

In converting one of these quantities to another, a conversion unit is available and is used "as is" or inverted, depending on the direction of the conversion (i.e., whether converting from grams to moles or from moles to grams).

Using H₂O as an example, it will be (or is) obvious that 1 gram of water is equivalent to the following.

$$1.0 \text{ g H}_2\text{O} \leftrightarrow 5.5 \times 10^{-2} \text{ moles H}_2\text{O} \leftrightarrow 3.3 \times 10^{22} \text{ molecules H}_2\text{O}$$

To accomplish these interconversions, the quantity is multiplied by the appropriate conversion factor, making use of the factor-label method. The conversion factor in each of these examples is underlined.

For example:

Given: 1.0 g H₂O   Convert to: moles H₂O

$$\frac{1.0 \text{ g H}_2\text{O}}{1} \cdot \frac{1 \text{ mole H}_2\text{O}}{18 \text{ g H}_2\text{O}} = 5.5 \times 10^{-2} \text{ moles H}_2\text{O}$$

Given: 1.0 g H₂O   Convert to: molecules H₂O

$$\frac{1.0 \text{ g H}_2\text{O}}{1} \cdot \frac{1 \text{ mole H}_2\text{O}}{18 \text{ g H}_2\text{O}} \cdot \frac{6.02 \times 10^{23} \text{ molecules}}{1 \text{ mole H}_2\text{O}} = 3.3 \times 10^{22} \text{ molecules H}_2\text{O}$$

Given: 5.5×10⁻² moles H₂O   Convert to: g H₂O

$$\frac{5.5 \times 10^{-2} \text{ moles H}_2\text{O}}{1} \cdot \frac{18 \text{ g}}{1 \text{ mole H}_2\text{O}} = 1.0 \text{ g H}_2\text{O}$$

Chapter 4 Chemical Equations

Given: 5.5×10^{-2} moles H_2O Convert to: molecules H_2O

$$\frac{5.5 \times 10^{-2} \text{ moles } H_2O}{1} \cdot \frac{6.02 \times 10^{23} \text{ molecules}}{1 \text{ mole } H_2O} = 3.3 \times 10^{22} \text{ molecules } H_2O$$

Given: 3.3×10^{22} molecules H_2O Convert to: moles H_2O

$$\frac{3.3 \times 10^{22} \text{ molecules } H_2O}{1} \cdot \frac{1 \text{ mole } H_2O}{6.02 \times 10^{23} \text{ molecules}} = 5.5 \times 10^{-2} \text{ moles } H_2O$$

Given: 3.3×10^{22} molecules H_2O Convert to: g H_2O

$$\frac{3.3 \times 10^{22} \text{ molecules } H_2O}{1} \cdot \frac{1 \text{ mole } H_2O}{6.02 \times 10^{23} \text{ molecules}} \cdot \frac{18 \text{ g } H_2O}{1 \text{ mole}} = 1.0 \text{ g } H_2O$$

As noted in the text, the conversion from grams to moles or visa versa requires only one step, while conversion between grams and molecules requires 2 steps or 2 conversion factors. The importance of using the factor-label method is brought out in these calculations as well.

Carry out the following conversions, indicating the conversion unit used in the factor-label analysis. In addition, the answer should have the correct number of significant figures.

A. GRAMS ↔ MOLES CONVERSION

Grams	Conversion Factor	Moles
(i) 4.0 g HF	_____	_____
(ii) 33 g KOH	_____	_____
(iii) 0.020 g AlF_3	_____	_____
(iv) _____	_____	0.30 moles PCl_3
(v) _____	_____	1.2 moles GeH_4
(vi) _____	_____	0.01 moles H_2CO_2

B. MOLES ↔ MOLECULES CONVERSION

Grams	Conversion Factor	Molecules
(i) 0.34 moles H_2	_____	_____
(ii) 11.3 moles Na_2SO_4	_____	_____

Chapter 4   Chemical Equations   33

    (iii) 3.4×10^{-24} moles NH_3 _____ _____

    (iv)   _____ _____ 1 molecule of I_2

    (v)   _____ _____ 6×10^{20} molecules O_3

    (vi)   _____ _____ 4×10^6 molecules $HClO_2$

C.   **GRAMS ↔ MOLECULES CONVERSION**

Grams	Conversion Factor	Molecules
(i) 0.14 g NCl_3	_____	_____
(ii) 4.0 mg S_8	_____	_____
(iii) 7.4 ug HI	_____	_____
(iv) _____	_____	4×10^{27} molecules $SeCl_2$
(v) _____	_____	6×10^{12} molecules $C_2H_4O_2$
(vi) _____	_____	2×10^2 molecules GeI_4

Exercise III.   Stoichiometry

In determining the weight relationships in chemical reactions, the key points are these:

1. The equation must be **balanced.**

2. **Molecules** react with **molecules**. **Moles** react with **moles**. The coefficients in the balanced equation dictate exactly how many of each of the reactants will react to form exactly how many product molecules. Therefore, the number of **moles** of the known reactant or product must be calculated first, if it is not given.

3. The coefficients in the balanced equation indicate the relative number of moles of each reactant and product. Therefore, multiply the moles of the known compound by the coefficient of the compound whose weight is unknown divided by the coefficient of the compound of known weight (or moles).

4. Convert the number of moles back to weight, by use of the appropriate conversion factor.

Consider the following equation. From the quantity given, determine the number of moles of the other quantities of interest.

$$C_7H_{16} + 11 O_2 \rightarrow 7 CO_2 + 8 H_2O$$

A.   Given: 0.1 mole C_7H_{16}

           moles O_2 moles CO_2 moles H_2O

Chapter 4 Chemical Equations   34

_____   _____   _____

B.   Given: 0.1 mole O_2

       _____ moles C_7H_{16} _____ moles CO_2 _____ moles H_2O

C.   Given 0.10 mole CO_2 produced

       _____ moles C_7H_{16} _____ moles O_2 _____ moles H_2O

SELF-TEST QUESTIONS

MULTIPLE CHOICE. In the following exercises, select the correct answer from the choices listed. In some cases, two or more answers will be correct.

1. The reaction of C_4H_{10} with O_2 to produce CO_2 and H_2O is what type of reaction?
   a. Dissociation
   b. Combustion
   c. Rusting
   d. Respiration

2. The formula weight for $Ca_3(PO_4)_2$ is:
   a. 279.3
   b. 310.3
   c. 167.3
   d. 543.3

3. A thimble of water contains 4.0×10^{21} molecules. The mass of water is:
   a. 1.1 g
   b. 11 g
   c. 1.1×10^{-8} g
   d. 0.11 g

4. Calculate the number of moles of aspirin, $C_9H_8O_4$, in a 4.0 g tablet.
   a. 2.3×10^{-4}
   b. 2.3
   c. 4.6×10^{-3}
   d. 0.023

5. The number of grams of K_2O produced from the oxidation of 4.0 grams of K is:
   a. 4.7 g
   b. 4.0 g
   c. 9.4 g
   d. 6.3 g

6. If the reaction in problem 5 actually produces only 1.0 g K_2O, what is the percentage yield for the reaction?
   a. 25%
   b. 11%
   c. 21%
   d. 16%

7. In the equation below, the oxidizing agent in the reaction is:
   $Al_{(s)} + H^+ = Al^{+3} + H_{2(g)}$
   a. $Al_{(s)}$
   b. H^+
   c. Al^{+3}
   d. H_2

8. The species which is oxidized in problem 7 is:
   a. $Al_{(s)}$
   b. H^+
   c. Al^{+3}
   d. H_2

Chapter 4   Chemical Equations   35

9. Indicate which reaction is the most endothermic.
    a.   A + B → C + 1,600 cal.
    b.   D + E + 12,000 cal. → F
    c.   G + H → I + 2.3 kcals.
    d.   X + Y + 2.6 kcal. → Z

10. A chemist used a laboratory calorimeter to determine that the combustion of glucose produced 686 kcal per mole of glucose. The amount of energy produced in the body by the metabolism of this same amount of glucose is:
    a.   Greater than 686 kcal.   c.   Equal to 686 kcal.
    b.   Less than 686 kcal.   d.   Can not predict

11. The correct answer to problem 10 derives from a law which states that the heat of reaction depends on the starting compounds and the final products and is independent of the reaction pathway. This law is called:
    a.   Law of Balanced Reactions   c.   Hess's Law
    b.   Avogadro's Law   d.   Hund's Rule

COMPLETION. Write the word, phrase or number in the blank which will complete the statement or answer the question.

1. The mass of germanium, Ge, (atomic number 32 and atomic weight 72.59) which contains a mole of Ge atoms is _____.

2. A _____ is used to determine the _____ value of foods.

3. The _____ _____ _____ is measured in a calorimeter. This is determined by measuring the temperature change of the water in the calorimeter. If the temperature decreases as a result of the reaction, the reaction is _____.

4. An _____ reaction is one in which _____ is given off, but not necessarily as _____.

5. _____ in organic compounds is defined as the gain of hydrogen or the loss of _____.

6. Four very common reaction types in aqueous solution which involve ions are (1) _____, (2) _____, (3) _____ or (4) _____.

7. Complete the following table.

Compound	Grams	Number of Moles	Molecules
a. NH_3	_____	6.0×10^{-4}	_____
b. $C_2H_2Br_2$	6.3	_____	_____
c. $Si_2C_6H_{18}$	_____	_____	9.0×10^{12}

8. Complete and balance the following equations.
    a. $C_5H_{12} + O_2 \rightarrow CO_2 + H_2O$

b. $Al^{+3} + SO_4^{2-} \rightarrow$

c. $SnO_2 + H_2 = Sn_{(s)} + H_2O$

d. $Au + Cl_2 \rightarrow AuCl_3$

e. $C_6H_{12}N_4 + H_2O \rightarrow H_2CO + NH_3$

f. $HBr + KOH \rightarrow KBr + H_2O$

9. Indicate the type of reaction which is written out in question 8, a-d and f.

10. Consider the reaction of X and Y to form Z, with the indicated stoichiometry.
    $$X + 2Y \rightarrow Z$$
    The formula weights are: X = 50; Y = 200; Z = 100

    a. The reaction indicates that _____ moles of X react with _____ moles of Y to produce _____ moles of Z.

    b. If 200 grams of X is consumed in the reaction, the number of grams of Y consumed is _____ and the number of moles of Z produced is _____.

11. Consider the same stoichiometric equation in question 9. If 3 moles of Z were formed in the reaction, then _____ grams of X reacted with _____ grams of Y.

ANSWERS FOR FOCUSED REVIEW EXERCISES

Exercise I.
- a. 223.2
- b. 62.0
- c. 204.0
- d. 295.5
- e. 48.0
- f. 284.0
- g. 89.0
- h. 300.0

Exercise II.

A. (i) 1 mole HF
       20 g HF ;   0.20 moles

   (ii) 1 mole KOH
        56.1 g KOH ;   0.59 moles

   (iii) 1 mole AlF$_3$
         84 g AlF$_3$;   2.4x10^{-4} moles

   (iv) 137.4 g PCl$_3$
        1 mole PCl$_3$;   41 g PCl$_3$

   (v) 76.6 g GeH$_4$
       1 mole GeH$_4$;   92 g GeH$_4$

   (vi) 46 g H$_2$CO$_2$
        1 mole H$_2$CO$_2$;   0.46 g H$_2$CO$_2$

Chapter 4   Chemical Equations   37

B. (i) $\underline{6.02 \times 10^{23} \text{ molecules}}$
           1 mole H$_2$; 2.0×10^{23} molecules

    (ii) $\underline{6.02 \times 10^{23} \text{ molecules}}$
           1 mole Na$_2$SO$_4$; 6.8×10^{24} molecules

    (iii) $\underline{6.02 \times 10^{23} \text{ molecules}}$
           1 mole NH$_3$;   2.0 molecules

    (iv) $\underline{\text{1 mole I}_2}$
           6.02×10^{23} molecules ; 4.2×10^{-22} g.

    (v) $\underline{\text{1 mole O}_3}$
           6.02×10^{23} molecules ; 0.48 g.

    (vi) $\underline{\text{1 mole HClO}_2}$
           6.02×10^{23} molecules ; 4.6×10^{-16} g.

C. (i) $\underline{\text{1 mole}}$ $\underline{6.02 \times 10^{23} \text{ molecules}}$
           120.4 g NCl$_3$ 1 mole   ; 7×10^{20} molecules

    (ii) $\underline{\text{1 mole}}$ $\underline{6.02 \times 10^{23} \text{ molecules}}$
           256.6 g S$_8$ 1 mole   ; 9×10^{18} molecules

    (iii) $\underline{\text{1 mole}}$ $\underline{6.02 \times 10^{23} \text{ molecules}}$
           127.9 g HI   1 mole   ; 3×10^{16} molecules

    (iv) $\underline{149.9 \text{ g SeCl}_2}$ $\underline{\text{1 mole}}$
           1 mole 6.02×10^{23} molecules ; 9.9×10^5 g

    (v) $\underline{60 \text{ g C}_2\text{H}_4\text{O}_2}$ $\underline{\text{1 mole}}$
           1 mole 6.02×10^{23} molecules ; 6×10^{-10} g

    (vi) $\underline{580.02 \text{ g GeI}_4}$ $\underline{\text{1 mole}}$
           1 mole 6.02×10^{23} molecules ; 1.9×10^{-19} g

Exercise III.
A. 1.1 moles O$_2$; 0.7 moles CO$_2$; 0.8 moles H$_2$O

B. 9×10^{-3} moles C$_7$H$_{16}$; 6×10^{-2} moles CO$_2$; 7×10^{-2} moles H$_2$O

C. 0.01 moles C$_7$H$_{16}$; 0.16 moles O$_2$; 0.11 moles H$_2$O

ANSWERS TO SELF-TEST QUESTIONS

Multiple Choice
1. b   5. a   9. b
2. b   6. c   10. c
3. d   7. b   11. c
4. d   8. a

Chapter 4 Chemical Equations   38

Completion
1. 72.59 g
2. calorimeter, caloric value
3. heat of reaction, endothermic
4. exergonic, energy, heat
5. reduction, oxygen
6. precipitation, gas formation, acid neutralization of a base, redox
7. a.   0.010 g, 3.6×10^{20} molecules
   b. 3.4×10^{-2} moles, 2.0×10^{22} molecules
   c.   2.2 ng, 1.5×10^{-11} moles
8. a. $C_5H_{12} + 8O_2 \rightarrow 5CO_2 + 6H_2O$
   b. $2Al^{+3} + 3SO_4^{-2} \rightarrow Al_2(SO_4)_3$
   c. $SnO_2 + 2H_2 = Sn_{(s)} + 2H_2O$
   d. $Au + 3/2 Cl_2 = AuCl_3$
   e. $C_6H_{12}N_4 + 6H_2O = 6H_2CO + 4NH_3$
9. a.   combustion
   b.   combination
   c.   single displacement
   d.   combination
   f.   double displacement
10. a.   one, two, one
    b.   1600 g, 4 moles
11. 150 g, 1200 g

Chapter 5   Gases, Liquids, and Solids

Happy is he who gets to know the
reasons for things.
   - Motto of Churchill College,
      Cambridge

5

Gases, Liquids and Solids

CHAPTER OBJECTIVES

After studying the chapter and working the assigned exercises in the text and study guide, you should be able to do the following.

1. State the basic assumptions in the kinetic-molecular theory of gases. Indicate how each individual assumption helps in understanding the behavior or properties of ideal gases.

2. Explain how a barometer and a manometer are setup to read gas pressure.

3. Name the four properties of a gas which defines its physical state and indicate the common units for each.

4. Apply Boyle's Law ($P_1V_1 = P_2V_2$, conditions of constant temperature) to calculate the final volume or pressure of a gas.

5. Use Charles's Law ($V_1/T_1 = V_2/T_2$, conditions of constant pressure) to calculate the final volume or temperature of a gas.

6. Utilize Gay-Lussac's Law ($P_1/T_1 = P_2/T_2$, conditions of constant volume) to calculate the final temperature or pressure of a gas.

7. Apply the combined gas law ($P_1V_1/T_1 = P_2V_2/T_2$) to calculate the final pressure, temperature or volume of a gas.

8. Use the ideal gas law ($PV = nRT$) to calculate either the pressure, volume, the number of moles or the temperature of the gas.

9. Utilize the modified form of the ideal gas law to calculate the molecular weight or the mass of a quantity of gas.

10. Utilize Avogadro's Law ($PV/RT = n$, at conditions of a specific pressure and temperature) to

Chapter 5 Gases, Liquids, and Solids

show that equal volumes of gases at the same temperature and pressure contain equal numbers of gas molecules.

11. Apply Dalton's Law of Partial Pressures, ($P_T = P_1+P_2+P_3...$), to calculate the partial pressure of a gas in a mixture or the total pressure in a mixture of gases.

12. Use Graham's Law of Diffusion, $[(R_A^2/R_B^2) = [M.W._B / M.W._A]$, to calculate either the relative rates of diffusion of two gases or the molecular weight of one of the gases.

13. Name, define and give the approximate energies involved in the three basic intermolecular forces between molecules.

14. Draw a few molecules which can participate in intermolecular hydrogen bonding and show the specific interactions involved.

15. Define boiling point and explain the factors which influence the temperature at which a liquid boils.

16. Indicate the specific make-up and the general characteristics of an a) ionic, b) molecular, c) metallic, d) polymeric, e) network and f) amorphous solid.

17. Draw a heating curve for ice, explain the main features of the curve and explain the significance of the heat of fusion and the heat of vaporization in this figure.

18. Define the important terms and comparisons in this chapter and give specific examples where appropriate.

19. Draw the phase diagram for water, pointing out the triple point and indicating the regions where only solid, liquid or gas exists.

Make a summary sheet of the important laws and corresponding equations presented in this chapter which provide the basis for characterizing the property of a gas. Indicate the parameters which are kept constant in each relationship and the units of each.

IMPORTANT TERMS AND COMPARISONS

Gaseous, Liquid and Solid State
Melting Point and Freezing Point
Barometer and Manometer
Torr, mm Hg, Pascal
Atmosphere (unit of pressure)
Systolic and Diastolic Pressure
Sphygmomanometer
Boyle's, Charles's and Gay-Lussac's Laws
22.4 L
Partial and Total Pressure
Dalton's and Graham's Laws
Phase Diagram and Triple Point
Entropy

Arterial and Venous Blood
London Dispersion Force and Hydrogen Bond
Fluidity
Vapor Pressure of a Liquid
Boiling Point and Melting Point
Crystallization
Crystals, Amorphous and Network Solids
Phase Change
Heating Curve
Heat of Fusion and Heat of Vaporization
Buckminsterfullerene or Bucky Balls

FOCUSED REVIEW

Chapter 5   Gases, Liquids, and Solids   41

One of the most perplexing difficulties that many students encounter in problem solving is deciding the equation to use to obtain the answer. Since there are a number of equations (relationships) which may be used in solving gas law problems, a simple procedure is necessary to find just the right relationship for the particular problem.

The parameters that may be involved in a particular problem are the following:

Parameter	Common Units
Pressure (P)	Atmosphere, mm Hg, Torr, Pascal
Volume (V)	Liter, milliliter
Temperature (T)	°C, °K
Moles (n)	--------
Mass (m)	Grams
Molecular Weight (M.W.)	Atomic Mass Units

The types of problems can be subdivided into two major categories. The first is the situation in which the gas is completely characterized in one state, the **conditions are CHANGED**, and one of the characteristics of the second state is unknown. A number of situations may arise.

TABLE I

	State 1		State 2	
	Known Quantity		Known Quantity	Unknown Quantity
Ideal Gas Law:	$\dfrac{P_1V_1}{T_1} = \dfrac{P_2V_2}{T_2}$			
	(P_1, V_1, T_1)	1. P_2V_2 2. P_2, T_2 3. V_2, T_2		T_2 or V_2 or P_2
Boyle's Law:	$P_1V_1 = P_2V_2$, at constant T			
	(P_1, V_1)	1. P_2 2. V_2		V_2 or P_2
Charles's Law:	$\dfrac{V_1}{T_1} = \dfrac{V_2}{T_2}$, at constant P			
	(V_1, T_1)	1. V_2 2. T_2		T_2 or V_2
Gay-Lussac's Law:	$\dfrac{P_1}{T_1} = \dfrac{P_2}{T_2}$, at constant V			
	(P_1, T_1)	1. P_2 2. T_2		T_2 or P_2

The second category of problems involves a gas or gases in one **PARTICULAR** state. **No changes in conditions are involved**. In this case, a specific property in this state is to be calculated, using either the ideal gas law, Dalton's Law of partial pressures, Avogadro's Law or Graham's Law of diffusion.

Chapter 5   Gases, Liquids, and Solids

Recall that the ideal gas law equation may be written in two forms.

$$PV = nRT \quad (1)$$

and since

$$n = \frac{g}{M.W.}$$

$$PV = \frac{gRT}{(M.W.)} \quad (2)$$

Therefore, one can solve for any one of the parameters in equation (1) or (2), if the other quantities are given in the problem.

Two final points of interest.

1. R, the universal gas constant is a physical quantity, which you'll recall has both a numerical value and **units**. The value of R presented in the text is R = 0.082 L-atm/mole-K. Therefore, it is essential in using this equation with this value of R that the gas parameters conform with these units of R. That is, the parameters must have units of P (atm), V (L), n (moles) and T (K). If the parameters are not given in the proper units, convert them immediately. Be on the outlook for just this situation.

2. In problems involving two states of a gas and using any one of the four gas laws presented, it is also essential that the units characterizing state 1 be the same as in state 2. For example, if the pressure in state 1 is given as 1.2 atm, while the pressure in state 2 is given as 500 mm Hg, one of these values (usually the 500 mm Hg value) must be converted to the other before the calculation.

Knowing these preliminaries and precautions, the procedure for analyzing and solving a gas law problem is:

1. Determine if the problem involves the gas in only one state or two different states.

2. Determine (**write out**) what is known and what is to be determined (unknown) in the problem.

3. From your analysis in steps 1 and 2, decide which law and equation is appropriate to solve the problem.

4. Rearrange the equation if necessary and check that all units are appropriate.

5. Solve for the unknown quantity.

To pull some of this together, consider the following problem.

If a sample of 1.78 moles of $CHCl_3$ gas occupies 68 L at a pressure of 275 mm Hg, what is its temperature in °C?

Step 1.   The problem involves $CHCl_3$ gas in **one state**. There are no changes to a new state.

Chapter 5 Gases, Liquids, and Solids

Therefore the equations as written in Table I (Study Guide) are not appropriate and either the (1) ideal gas law, (2) Dalton's Law of partial pressures, (3) Avogadro' Law or (4) Graham's Law of diffusion will be used.

Step 2.

Knowns	Unknown
n = 1.78 moles	
V = 68 L	T = ?
P = 275 mm Hg	

Step 3. From the parameters given and the unknown quantity, the ideal gas law should be used in the form.

$$PV = nRT$$

Step 4. Rearrange by dividing through each side by nR. This yields

$$T = \frac{PV}{nR}$$

Convert pressure (P = 275 mm Hg) into units of atmosphere to conform with units of R, the universal gas constant.

$$P = \frac{275 \text{ mm Hg}}{1} \cdot \frac{1 \text{ atm}}{760 \text{ mm Hg}} = 0.362 \text{ atm}$$

Step 5. $$T = \frac{(0.362 \text{ atm})(68 \text{ L})}{(1.78 \text{ moles})(0.082 \text{ L-atm/mole-K})} = 169 \text{ K}$$

The final step in this problem is to convert the temperature from K to °C.
°C = 273° + K = 273° + 169,

Therefore, the final answer is: **T = 442°C**

SELF-TEST QUESTIONS

MULTIPLE CHOICE. Select the correct answer from the choices given. Only one of the choices is correct.

1. A closed aerosol-can which had been filled with shaving cream contains 3.0 g of nitrogen gas (N_2) as the propellant. If the volume of the can as 200 mL, the pressure in the can at 25°C is:
   a. 1.1 atm
   b. 130 atm
   c. 1.3 atm
   d. 13 atm

2. If the can in question 1 is thrown into an incinerator or fire, the temperature in the can could well get as high as 850 K. The pressure in the can, assuming that it did not explode, would be:
   a. 37 atm
   b. 3.7 atm
   c. 370 atm
   d. 370 mm Hg

3. A gas, which is initially contained in a movable cylinder of 900 mL volume and at 25°C., exerts a pressure of 3 atmospheres. The pressure is reduced at constant temperature to a value of 380 mm Hg. The new volume is:
   a. 5400 L
   b. 5.4 L
   c. 7.1×10^{-3} L
   d. 7.1 L

Chapter 5 Gases, Liquids, and Solids   44

4. A piece of Dry Ice (solid carbon dioxide, CO_2) is placed in a 1 L tank. It sublimes to CO_2 gas at STP. The amount of CO_2 was:
    a. 1 mg
    b. 44 g
    c. 4.4 g
    d. 2 g

5. The average kinetic energy of molecules is proportional to the
    a. temperature in °C
    b. velocity of the molecule
    c. temperature in K
    d. volume at STP

6. The official SI unit of pressure is:
    a. torr
    b. pascal
    c. atmosphere
    d. mm Hg

7. The autoclave operates on the principle of:
    a. Dalton's Law of Partial Pressure
    b. Boyle's Law
    c. Gay-Lussac's Law
    d. Charles's Law

8. A gas at 25°C is initially contained in a movable cylinder of 2000 mL volume and exerts a pressure of 38 mm Hg. The pressure is changed to 0.5 atm and the volume of the cylinder increases to 2.2 L. The final temperature in the cylinder is:
    a. 3×10^3 K
    b. 275 K
    c. 3.2 K
    d. 2×10^4 K

9. Consider the reaction: $3K + Y \rightarrow 2Z$. At STP, 2 L of Z are produced from the reaction of X and Y. The number of liters of X necessary is:
    a. 1 L
    b. 2 L
    c. 3 L
    d. cannot determine

10. A quantity of 6 grams of a gas occupies 300 milliliters at STP. The molecular weight of the gas is:
    a. 440
    b. 660
    c. 44
    d. 66

11. Predict which would exhibit the greatest surface tension.
    a. CCl_3H
    b. H_2O
    c. CH_3OH
    d. CCl_4

12. Indicate which of the following are characteristics of "buckyballs". There may be more than one answer.
    a. Contains 60 carbon atoms
    b. Carbons are arranged in pentagons and octagons
    c. Contains only carbon atoms
    d. Each carbon is bonded to four other carbons

13. Which are isoforms of carbon. There may be more than one answer.
    a. diamond
    b. soot
    c. graphite
    d. fullerenes

COMPLETION. Write the word, phrase or number in the blank which will complete the statement or answer the question.

Chapter 5   Gases, Liquids, and Solids

1. STP indicates conditions in which T = _____ and P = _____.

2. The boiling point of HF (M.W. = 18) is higher than that for CH_4 (M.W. = 18) because of _____, while the boiling point of n-octane is higher than that for propane because of the _____.

3. The behavior of an ideal gas at constant volume but at varying pressure or temperature is governed by _____ Law.

4. The O_2 gas in a 1 L tank at 1 atmosphere is transferred to a 2 L tank which already contains 2 atmospheres of N_2 gas. The final pressure is _____.

5. The higher the temperature, the _____ the _____ energy of molecules. Therefore, by heating a closed container of gas, the pressure will _____.

6. A 5.6 L container of Ne or O_2 contains _____ moles of gas or _____ molecules at STP.

7. The energy required to break a hydrogen bond is approximately _____ kcal/mole.

8. A molecule used in perfume is determined to diffuse one-tenth as fast as neon gas. Its molecule weight is _____.

9. The basic postulates in the kinetic molecular theory of gases are:
   i. _____

   ii. _____

   iii. _____

   iv. _____

   v. _____

   vi. _____

10. The melting point of diamond, which is a _____, is much higher than that for a _____ crystal.

11. The two plateaus in a heating curve represent the temperatures at which two phases _____. The amount of heat necessary to convert one gram of water at 100°C to steam is called the _____.

12. The intermolecular forces responsible for the solidification of Xe and other noble gases are _____.

13. The unit of pressure, _____, is equivalent to the _____.

14. Convert the physical quantities in the left column to their equivalent in the right-hand column.
    a. 4.3 mL   _____ ____ L

    b. 34°C   _____ K

Chapter 5   Gases, Liquids, and Solids   46

    c.   450 mm Hg   _____ atm

    d.   43 g O_2 _____ moles O_2

    e.   372 K   _____ ºC

    f.   STP   _____ mm Hg

                                                          _____ ºC

    g.   17 cal/mole   _____ Kcal/mole

15.   Match up the unit or instrument in the column at the right with its use in the left-hand column.

              Unit or Instrument   Use
     a.   Barometer   i.   Measure blood pressure
     b.   Sphygmomanometer   ii.   Laboratory instrument to measure pressure
     c.   Closed container of liquid   iii.   Hospital unit used for patients with CO
     d.   Hyperbaric Chamber   poisoning, smoke inhalation, in addition to
     e.   Manometer   other uses
                                                 iv.   Measures atmospheric pressure
                                                 v.   Used in determination of vapor pressure

16.   Water has a superstructure involving intermolecular _____, which in part explains why the density of _____ is greater than that of _____.

17.   As the temperature decreases, the vapor pressure of a liquid _____.

18.   _____ is the transition from the solid state directly into the _____ state without going through the _____ state.

19.   An amorphous solid differs from a crystalline solid in that the molecules or ions are solidified in a _____ pattern.

20.   Write the equation for the behavior of an ideal gas. _____.

21.   A new form of carbon, which contains ___ carbons and has a structure which resembles a soccer ball is called either _____ or _____.

22.   A measure of order is called _____. As the temperature of a substance increases in temperature, its _____ value will _____.

23.   Examples of solids that can be described as molecular, polymeric or network are _____, _____, and _____, respectively.

24.   Two types of solids that exhibit very high melting points are _____ and _____. Sodium chloride (NaCl) is an example of a _____ solid.

25.   A phase diagram plots _____ on the x axis and _____ on the y axis.

26.   The line in a phase diagram that separates the solid phase from the liquid phase contains all the _____ points for that substance.

Chapter 5 Gases, Liquids, and Solids 47

27. The _____ _____ is a unique condition in which all three phases coexist.

ANSWERS TO SELF-TEST QUESTIONS

Multiple Choice
1. d
2. a
3. b
4. d
5. c
6. b
7. c
8. a
9. c
10. a
11. b
12. a, c
13. a, b, c, d

Completion
1. T = 0°C, P = 1 atm
2. intermolecular hydrogen bonds, (higher) molecular weight
3. Gay-Lussac's Law
4. 2.5 atm
5. greater, kinetic, increase
6. 0.25, 1.5×10^{23}
7. 5-10
8. 2000
9. These statements can be paraphrased.
   i. A gas consists of molecules traveling in random directions, in straight lines and at a range of speeds.
   ii. The average kinetic energy, and therefore the average velocity of gas molecules, is directly proportional to the absolute temperature.
   iii. Gas molecules collide with each other and may exchange energy between each other, but the total kinetic energy is conserved.
   iv. The molecules in a gas take up essentially no volume. The volume of the gas is the volume of the container it is in.
   v. There are no attractions between the gas molecules.
   vi. The pressure of the gas is due to collisions of the molecules with the walls of the container. The greater the number of collisions per unit time, the greater the pressure.
10. network solid, molecular
11. coexist, heat of vaporization
12. London dispersion forces
13. torr, mm Hg or also 1 atm = 101,325 pascals is acceptable. The former equivalence is more commonly used, however.
14. a. 4.3×10^{-3} L
    b. 307 K
    c. 0.592 atm
    d. 1.3 moles O_2
    e. -1°C
    f. 760 torr, 0°C
    g. 1.7×10^{-2} kcal/mole
15. a. iv
    b. i
    c. v
    d. iii
    e. ii
16. hydrogen bonding, liquid water, ice
17. decreases

Chapter 5 Gases, Liquids, and Solids

18. sublimation, gaseous, liquid
19. random
20. PV = nRT
21. 60; buckminsterfullerene; bucky balls
22. entropy, entropy, increase
23. sugar or ice; rubber, plastic or proteins; diamond or quartz
24. ionic, network, ionic
25. T, P
26. melting
27. triple point

Chapter 6 Solutions and Colloids

The fluids of the human body and the seas of the
earth contain common table salt, sodium chloride,
and an assortment of other chemicals.
- S. Brooks
  The Seas Inside Us: Water in the Life Processes

6

Solutions and Colloids

CHAPTER OBJECTIVES

After studying the chapter and working the assigned exercises in the text and study guide, you should be able to do the following.

1. Summarize the characteristics of a mixture, a suspension, a colloidal dispersion and a true solution and give specific examples of each.

2. Indicate the factors which influence the solubility of a (1) gas, (2) liquid or (3) solid in a liquid solvent.

3. Perform calculations to determine the concentration of a solution expressed in (1) molarity, (2) % w/v and (3) w/w. Given the concentration of a solution in any of these forms, convert to its equivalent in the other two.

4. Using the formula, moles = $M_1 \times V_1 = M_2 \times V_2$, calculate the concentration of solutions made by the dilution of a more concentrated solution.

5. Outline the characteristics of water which are most responsible for the unusual properties, such as (1) its high boiling point and (2) its ability to dissolve ionic compounds.

6. Specify the fundamental difference between a strong and weak electrolyte and a non-electrolyte.

7. Discuss the mechanisms by which water can effectively dissolve particular covalent molecules.

8. Considering both molecular and ionic compounds, carry out calculations to determine the freezing point depression produced by the various aqueous solutions.

9. Define osmosis and osmolarity and discuss how these concepts relate to specific areas of

Chapter 6 Solutions and Colloids

medicine.

10. Define the important terms and comparisons in this chapter and give specific examples where appropriate.

IMPORTANT TERMS AND COMPARISONS

Homogeneous and Heterogeneous Mixtures
Solute, Solvent and Solution
Clear vs. Colorless
A Physical Constant
Saturated, Unsaturated and Supersaturated Solutions
"Like Dissolves Like"
Thermal Pollution
Concentration
Dilute vs. Concentrated
Percent Concentration
Hemodialysis
(% w/v), (w/w) and (v/v)
Molarity and Osmolarity

Proof
Solvation
Plaster of Paris
Electrolyte and Non-Electrolyte
Strong and Weak Electrolyte
Solution and Suspension
Transparent, Translucent and Opaque
Solvation Layer
Physiological Saline
Isotonic, Hypertonic and Hypotonic
Hemolysis and Crenation   Dialysis and

Reverse Dialysis

FOCUSED REVIEW OF CONCEPTS

In all chemically related fields, we are continuously called upon to make observations or measurements and to characterize a mixture or solution, either in a qualitative or quantitative sense. Recall that chemistry is an experimental science and our current knowledge is derived from **careful observations** of experimental happenings. Concepts in this chapter lay the foundation for such characterization. Consider, for example, that you are presented with a test tube containing some liquid and asked to characterize it as completely as you can. Your preliminary, yet very important, judgments will necessarily be **qualitative**. Initially, you might focus on these questions.

1. Does the vial contain a true solution, a colloidal dispersion or a suspension?

2. If the vial contains a true solution,
    a. Is the solution clear (transparent), translucent or opaque?
    b. Is the solution colored or colorless?
    c. Does the solution conduct electricity?
    d. Is the liquid simply a solvent or an actual solution?
    e. If it is a solution, is it dilute or concentrated and is it unsaturated, saturated or supersaturated with solute?

At this point, it should be evident that although the characteristics are qualitative, we know a great deal about the unknown solution in the test tube.

To characterize the solution further, however, requires **quantitation**. The following table collects a number of means of quantitating solutions, with the definition and examples of each. However, first just focus your attention on a basic point which is often times not clearly defined and leads to difficulty. You will note that all the quantitative concentration terms in Table I are expressed as the quantity of one reagent in a certain volume of **SOLUTION** (= solute plus solvent). The emphasis here is to distinguish the difference between the volume of the total solution from the volume of the solvent used to dissolve the solute.

Chapter 6 Solutions and Colloids

TABLE I
Quantitative Description of Solutions

Concentration Term	Definition	Examples
Molarity (M)	Moles solute per 1 L **SOLUTION** $M = (moles/L)$	3 M NaCl; Dissolve 175 g NaCl in enough water to make 1 L solution.
Osmolarity	Molarity times the number of particles (n) produced by each mole of solute Osmolarity = n x M (n = 5)	0.1 M NaCl; Osmolarity = 0.2 NaCl → Na⁺ + Cl⁻; n = 2 0.3 M Mg₃(PO₄)₂; Osmolarity = 1.5 0.1 M Mg₃(PO₄)₂ → 3Mg⁺² + 2PO₄³⁻; Osmolarity = 0.5 0.8 M glucose (n = 1); Osmolarity = 0.8
(% w/v)	Weight solute per 100 mL **SOLUTION**	5 g of X in 100 mL solution 5 % w/v
(% w/w)	Weight solute per weight **SOLUTION**	5 g of X in 50 g solution 10 % w/w
(% v/v)	In a 2 component solution, volume of A added to enough liquid B to make 100 mL **SOLUTION**	17 mL liquid A plus enough liquid B to make 100 mL solution 17 % v/v

SELF-TEST QUESTIONS

MULTIPLE CHOICE. In the following exercises, select the correct answer from the choices listed. In some cases, two or more choices will be correct.

1. Which of the following gases can dissolve in rain droplets to produce acid rain.
   a. O_2
   b. SO_3
   c. CO_2
   d. CO

2. The solubility of a solid solute in water will depend on what factors?
   a. temperature
   b. size of the solute particles
   c. polarity of solute
   d. pressure

3. Which of the following covalent molecules can hydrogen-bond to water?
   a. nitrogen, N_2
   b. ammonia, NH_3
   c. methane, CH_4
   d. ethyl alcohol, C_2H_5OH

4. Indicate which of the following properties a colloidal dispersion has that a true solution does not.
   a. homogeneity
   b. Tyndall effect
   c. filterable with ordinary paper
   d. electrical conductance

5. An example of a colloid in which a solid is dispersed in a liquid is:
   a. cheese
   c. jellies

Chapter 6 Solutions and Colloids   52

      b.   whipped cream   d.   clouds

6. Which of the following properties does a suspension **not** have?
   a. settles on standing
   b. homogeneous
   c. translucent or opaque
   d. particles size of < 1000 nm

7. A solution containing proteins and NaCl salt is placed in a semipermeable membrane bag and dialyzed against distilled water. Which component or components will be found in the distilled water?
   a. NaCl, proteins
   b. NaCl
   c. proteins
   d. none of the components

8. The solubility of a polar gas molecule in a liquid solvent increases with:
   a. increasing temperature
   b. decreasing temperature
   c. increased gas pressure
   d. increased solvent polarity

9. The preparation of 300 mL of a 2.0×10^{-5} M solution of hemoglobin (molecular weight 64,000) requires how many grams of hemoglobin?
   a. 0.38 g
   b. 38 g
   c. 6.4×10^{-1} g
   d. 7.6 mg

10. The reaction of oxygen gas with hemoglobin (Hb) is shown in the equation below:
    $$Hb + 4O_2 = Hb(O_2)_4$$

    How many moles of O_2 can react with the hemoglobin solution in problem 9?
    a. 3×10^{-2} moles
    b. 2.4×10^{-5} moles
    c. 6×10^{-6} moles
    d. 3×10^{-6} moles

11. A solution of HCl is prepared to be 1×10^{-2} M. A 15 mL aliquot is taken and diluted with 80 mL of water. The final concentration of HCl is
    a. 1.9×10^{-4} M
    b. 1.9×10^{-3} M
    c. 1.6×10^{-3} M
    d. 1.6×10^{-4} M

12. A 1.5 M solution is made by dissolving 50 g of compound X in 50 mL of solution. The molecular weight of X is :
    a. 660
    b. 100
    c. 150
    d. 300

13. To make a dye solution for coloring an old white shirt, a student dissolves 5 mg of dye to 10 L of water. Calculate the concentration of this solution in ppb. (Recall that 1 mL of water weighs 1 g).
    a. 5000 ppb
    b. 50 ppb
    c. 500 ppb
    d. 5 ppb

COMPLETION. Write the word, phrase or number in the blank space or draw the appropriate structure in answering the question.

1. Preparation of 500 mL of a 0.5 M NaCl solution requires _____ NaCl.

2. The osmolarity of a 0.1 M $Na_3(PO_4)$ solution is _____. The freezing point of this solution is determined on a 50 mL portion (aliquot) of the solution. The observed freezing

Chapter 6 Solutions and Colloids   53

   point will be _____.

3. Crystalline substances that contain no water are called _____. Substances which become hydrated on exposure to the open air are referred to as _____ compounds.

4. Three 0.1 M solutions are prepared, one with HF, another with NaCl and the last one with sucrose (common sugar). Because the labels came off the bottles, we shall identify the solutions as simply solutions A, B and C. Solutions marked A and B conduct electricity; however, solution A clearly has a much greater capacity to conduct electricity than does solution B. Solution C does not conduct electricity. Identify each solution and characterize it according to its electrolytic character.

Compound	Solution	Type of Electrolyte
HF	(i)	(iv)
NaCl	(ii)	(v)
Sucrose	(iii)	(vi)

5. A true solution contains particles with a maximum diameter of _____.

6. A _____ solution has a greater osmolarity than do red blood cells. Therefore, placing red blood cells in this solution will cause the cells to _____. A 0.3 M glucose solution is a _____ (tonic) solution.

7. The concentration of a 0.40 M glucose, $C_6H_{12}O_6$, solution, can be expressed in terms of (% w/v). The equivalent concentration expressed in this way would be _____ (% w/v).

8. The random motion of a colloid suspended in a solvent is called _____.

9. A biochemist prepares a solution containing hemoglobin, the compound (a protein) in the red blood cells that carries O_2 to our cells. The hemoglobin in this solution is referred to as the _____.

10. A student adds benzene to a bottle containing CCl_4 and also to a separate bottle containing methyl alcohol, CH_3OH. The benzene is miscible in the _____, but immiscible in the _____. This is an example of the rule that "like dissolves like." In this case, a _____ compound dissolves in a _____ solvent.

11. Osmosis involves the passage of _____ from the _____ to the _____ side of a semipermeable membrane.

12. _____ is the process that is used in hospitals to remove waste products from the blood of patients who have kidneys that are not functioning properly.

ANSWERS TO SELF-TEST QUESTIONS

Multiple Choice
1. b, c
2. a, c
3. b, d
4. b
5. c
6. b, d
7. b
8. b, c, d
9. a
10. b
11. c
12. a
13. c

Chapter 6 Solutions and Colloids

Completion
1. 14.6 g
2. 0.4; -0.76°C
3. anhydrous; hygroscopic
4. (i)   Solution B   (iv) weak electrolyte
   (ii) Solution A   (v)   strong electrolyte
   (iii) Solution C   (vi) non-electrolyte
5. 1 nm
6. hypertonic; shrivel; isotonic
7. 7.2 (% w/v)
8. Brownian motion
9. solute
10. CCl_4; CH_3OH; non-polar; non-polar
11. solvent (molecules); dilute; (more) concentrated
12. Hemodialysis

Chapter 7 Reaction Rates and Chemica Equilibrium

Hofstadter's Law:
It always takes longer than you expect, even when you take into account Hofstadter's Law.
- Douglas Hofstadter

7

Reaction Rates and Equilibrium

CHAPTER OBJECTIVES

After you have studied the chapter and worked the assigned exercises in the text and the study guide, you should be able to do the following.

1. Given the concentration of the reactants and/or products at various times during the reaction, calculate the rate of reaction.

2. Distinguish between molecular collisions and effective molecular collisions and how these relate to the rate of a reaction.

3. Clarify how effective collisions, temperature and activation energy relate to the rate of reaction.

4. Draw and label completely the energy diagrams for (i) an endothermic reaction and (ii) an exothermic reaction.

5. Describe how a catalyst influences a reaction and its affect on the energy diagram.

6. Define the composition of the activated complex and its position in the energy diagram.

7. Discuss how each of the following factors affects the rate of reaction: (1) nature of reactants, (2) concentration of reactants and products, (3) temperature and (4) the presence of a catalyst.

8. Describe the concept of dynamic equilibrium as it relates to a reversible reaction.

9. Given the equilibrium expression, write out the chemical reaction.

10. Given the chemical reaction, write out the equilibrium expression.

11. Given the concentration of the reactants and products at equilibrium, calculate the equilibrium constant, K.

Chapter 7   Reaction Rates and Chemical Equilibrium

12. Use LeChatelier's Principle to explain how the following factors influence the equilibrium: (1) addition of a reaction component; (2) removal of a reaction component; (3) addition of a catalyst.

13. Utilizing LeChatelier's Principle and considering both an exothermic and endothermic reaction, explain how a change in temperature (increase or decrease) will alter the equilibrium constant.

14. Define the important terms and comparisons in this chapter and give specific examples where appropriate.

IMPORTANT TERMS AND COMPARISONS

Chemical Kinetics
Reaction Rate
Effective and Ineffective Collision
Activation Energy
Transition State and Activated Complex
Energy Diagram
Endothermic and Exothermic Reactions
Endergonic and Exergonic Reactions

Catalyst and Enzyme
Reversible and Irreversible Reaction
Dynamic Equilibrium
Haber Process
Fixation of Nitrogen
Equilibrium Constant and Equilibrium Expression
LeChatelier's Principle

SELF-TEST QUESTIONS

MULTIPLE CHOICE. In the following exercises, select the correct answer from the choices listed. In some cases, more than one choice will be correct.

1. Consider the reaction, A + B → C. It was found that at 18 minutes after the start of the reaction, the concentration of C was 0.04 M. The rate of this reaction is:
   a.   450 M/min
   b. 4.2×10^{-3} M/min
   c.   900 mole/L-min
   d. 2.1×10^{-3} mole/L-min

2. Molecular collisions in the reaction, A + B → C, are necessary for a reaction to take place because:
   a.   Reactant molecules must directly interact with each other to produce product.
   b.   The collisions of rapidly moving molecules provide energy.
   c.   Both reasons above.

3. The activated complex in the reaction A + B → C + D is composed of:
   a.   A and B
   b.   C and D
   c.   A
   d.   A and D

4. Reaction X + Y → Z is completed in 2 hours at 20°C. The reaction is carried out at a different temperature at which the reaction is complete in 7.5 minutes. This new temperature is:
   a.   293 K
   b.   40°C
   c.   333 K
   d.   30°C

5. The equilibrium constant is equal to 5×10^{-4} for the reaction, A + B + C = 2D. The equilibrium expression for the reaction is:

Chapter 7   Reaction Rates and Chemical Equilibrium

a. $K = \dfrac{(A)(B)(C)}{(D)^2}$

b. $K = \dfrac{(2D)}{(A)(B)(C)}$

c. $K = \dfrac{(D)^2}{(A)(B)(C)}$

d. $K = \dfrac{(2D)^2}{(A)(B)(C)}$

6. If the equilibrium concentrations for the reactants and products in the reaction in question 5 are $[A] = [B] = [C] = 1\times10^{-3}$ M, the concentration of D would be:
   a. 5×10^{-10} M
   b. 5×10^{-13} M
   c. 7×10^{-7} M
   d. 5×10^{-5} M

7. Consider the reaction, $Cl_{2(g)} + H_2O_{(l)} = H^+ + Cl^- + HOCl$. The molarity values for of the chemical species at equilibrium are determined to be: $[Cl_{2(g)}] = 0.10$; $[H_2O] = 0.24$; $[H^+] = 1\times10^{-3} = [Cl^-]$; $[HOCl] = 0.4$. The equilibrium constant for this reaction is:
   a. 1.7×10^{-5}
   b. 1.7×10^{-2}
   c. 6×10^4
   d. 6×10^2

8. Which of the following describe the Haber process?
   a. $N_{2(g)} + 3H_{2(g)} = 2NH_{3(g)}$
   b. Exothermic
   c. Uses a catalyst
   d. All of the above

9. If one monitors the rate of the formation of C in the reaction, $A + B \rightarrow C$, one finds the rate is 5×10^{-2} M/min. After one hour of reaction, the concentration of C is:
   a. 5×10^{-2} M
   b. 3 M
   c. 8×10^{-4} M
   d. 30 M

COMPLETION. Write the word, phrase or number in the blank space that completes the statement.

1. Indicate whether the following reactions are endothermic or exothermic.
   a. A + B + Energy → C   _____
   b. A + B → C + D + Energy   _____
   c. A + B → C $\Delta E = -40$ kcal/mole   _____
   d. $H_{2(g)} + O_{2(g)} \rightarrow H_2O_{(e)}$ + Energy   _____

2. Indicate how each of the following will affect the rate of formation of C in the reaction, A + B = C. The answer should be increase, decrease or no effect (on rate).
   a. Add a catalyst   _____
   b. Increase concentration of A   _____
   c. Decrease concentration of B   _____
   d. Decrease the temperature   _____

3. Consider the exothermic reaction, A + B → C + D, -10 kcal, in which the activation energy for the reaction is 25 kcal. Complete the diagram below with labels and showing all items of importance.

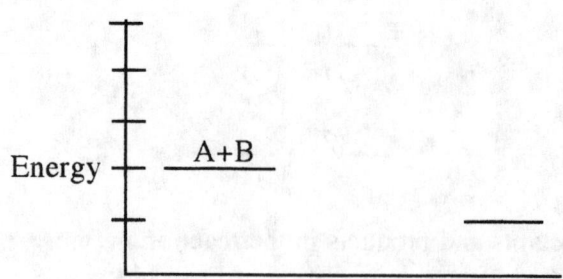

4. Completely label the following diagram by placing the appropriate label in each indicated block.

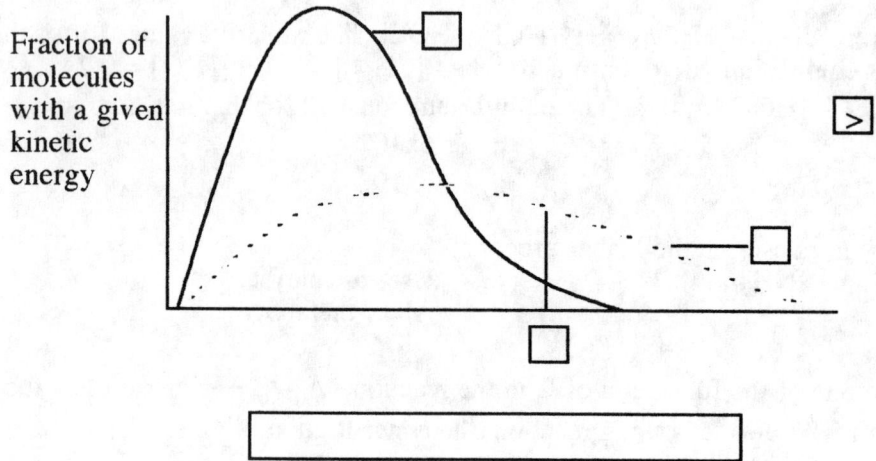

5. Molecules that act as catalysts in the human body are called _____.

6. Catalysts in a reaction lower the _____ of the reaction by increasing the number of _____ between reactant molecules without increasing the reaction temperature. However, the energy of the reactants and products (do or do not) _____ change by the presence of a catalyst.

7. Complete the following table.

   Equilibrium Expression   Reaction
   a. $H_2 + O_2 = H_2O$

   b. $\dfrac{(W)^2}{(A)(B)} = K$

   c. $H_2O = H^+ + OH$

8. The concept of a dynamic equilibrium brings out the finding that although the _____ of both reactant and product molecules (does or does not) change, both the _____ and _____ reactions are taking place.

9. Using LeChatelier's Principle, indicate how the following changes will affect the equilibrium in the exothermic reaction. Indicate whether the reaction goes to the right or to the left or has no change.

   $$A + B = C$$

   Change   Effect
   a. Increase [A]

b. Decrease [C]
c. Add a catalyst
d. Increase temperature

ANSWERS TO SELF-TEST QUESTIONS

Multiple Choice
1. d
2. c
3. a and b; note that A and B are the only reactants and C and D are the only products. The atoms in A and B are the same ones in C and D, only rearranged to form the new molecules.
4. c
5. c
6. c
7. a
8. d
9. b

Completion
1. a. endothermic
   b. exothermic
   c. exothermic
   d. exothermic
2. a. increase rate
   b. increase rate
   c. decrease rate
   d. decrease rate (number of effective collisions is reduced

3.

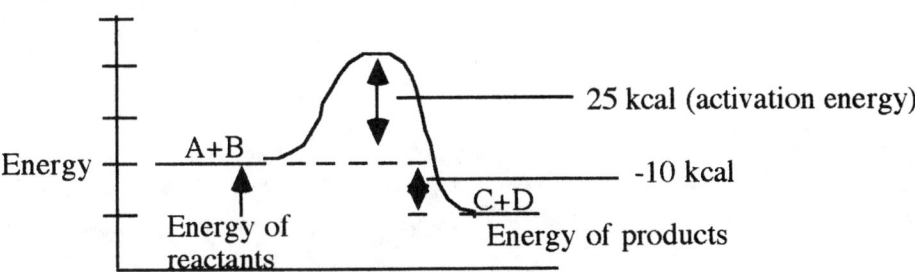

4.

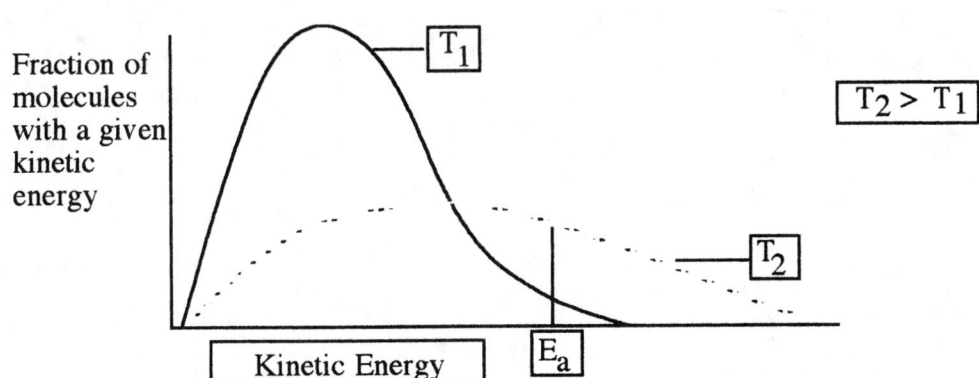

5. enzymes
6. activation energy, effective collisions, do not
7. a. $K = \dfrac{(H_2O)}{(H_2)(O_2)}$

b. A + B = 2 W

c. $K = \dfrac{(H^+)\,(OH^-)}{(H_2O)}$

8. concentrations, does not, forward, reverse
9. a. reaction goes to right
   b. reaction goes to right
   c. no effect
   d. reaction goes to left. In an exothermic reaction, heat can be considered a product. Therefore, increasing the temperature will increase the heat (a product) and the reaction will go to the left to relieve the stress.

Concerning solutions to acid-base
equilibrium problems: The answers
to the riddle of life are not found
in the back of the book.
 - *Charlie* Brown

8

Acids and Bases

CHAPTER OBJECTIVES

As will become evident in this chapter and those focused on biochemistry (Chapters 16-26), the body contains aqueous systems vital to life as we know it. These systems include intracellular fluids, blood, gastric juices, saliva and more. The importance of understanding the **CHARACTERISTICS AND REACTIONS** of **ACIDS**, **BASES** and **SALTS** in aqueous solutions cannot therefore be overemphasized. This chapter dwells on these fundamentals.

After you have studied the chapter and worked the assigned exercises in the text and the study guide, you should be able to do the following.

1. Characterize Arrhenius acids and bases and define the similarities and differences with Bronsted-Lowry acids and bases. Give specific examples of each.

2. Explain the difference between strong and weak acids (bases) and give four examples of strong acids (bases).

3. Write the reactions involving Bronsted-Lowry acids and bases and point out the conjugate acid-base pairs.

4. Define Ka and pKa values and explain how they are related to acid strength.

5. Write ionic equations for the neutralization reaction of acids and bases; clearly define the salt which forms in each case.

6. For an aqueous solution, define the neutral point and the acidic and basic domains in terms of pH, pOH, [H$^+$] and [OH$^-$]. At any pH, show the relationship between pH and pOH and also [H$^+$] and [OH$^-$].

7. Given the [H$^+$] or [OH$^-$] of a solution, calculate the pH. Given the pH or pOH of a solution,

Chapter 8 Acids and Bases

calculate the [H$^+$].

8. Name six body fluids or common materials which are (1) acidic or which are (ii) basic. In each case, indicate its approximate pH value.

9. Explain the importance of acid-base indicators in qualitatively testing the acidic or basic character of a solution.

10. Make a list of salts and characterize them as either neutral, acidic or basic salts.

11. Write equations for the hydrolysis reactions associated with acidic and basic salts. Indicate whether the resulting solution is acidic or basic.

12. Define buffer solution and indicate the components and their relative concentrations necessary to prepare buffers with good buffering capacity.

13. Show how the components of carbonate and phosphate buffers react on addition of strong acid or base.

14. Explain in some detail the way in which you would determine the concentration of a basic solution by titration with a strong acid of known concentration.

15. Name a few acids and bases in which the equivalent weight is (i) the same as the formula weight, (ii) different than the formula weight. Explain the reason for this difference.

16. For the same acids and bases as in objective 15, indicate in which cases the normality of the solution is equal to or greater than the molarity.

17. Using the formula, $V_{acid} \times N_{acid} = V_{base} \times N_{base}$, and the data from a titration, calculate the normality of an acidic or basic solution of unknown concentration.

18. Define the important terms and comparisons in this chapter and give specific examples where appropriate.

IMPORTANT TERMS AND COMPARISONS

Arrhenius Acid and Base
Hydronium Ion and Hydroxide Ion
Strong and Weak Acid
Glacial Acetic Acid
Bronsted-Lowry Acid and Base
Conjugate Acid-Base Pair
Mono-, Di- and Triprotic Acid
Acidity Constant, Ka
pKa, pH and pOH
Ionic Equation
Antacid
Metal Oxide and Hydroxide
Baking Soda
Molarity and Normality

Ion Product of Water, Kw
Acidic and Basic Solution
pH Indicator
pH Paper
pH Meter
Hydrolysis
Basic, Acidic or Neutral Salts
Buffering Action
Acidosis and Alkalosis
Hyper- and Hypoventilation
Buret
Pipet
Molecular Weight and Equivalent Weight

Chapter 8   Acids and Bases

SELF-TEST QUESTIONS

MULTIPLE CHOICE. In the following exercises, select the correct answer from the choices listed. In some cases, more than one choice will be correct.

1. The pH of a 1 x 10^{-4} M NaOH solution is
   a. 4
   b. 1
   c. 10
   d. 6

2. Which of the following is (are) strong acid(s)?
   a. perchloric acid
   b. boric acid
   c. hydroiodic acid
   d. acetic acid

3. Which of the following is both a strong acid and an oxidizing agent?
   a. NH_3
   b. H_2SO_4
   c. HNO_3
   d. H_2O

4. Indicate which of the following characteristics are associated with a strong acid solution.
   a. concentrated
   b. low pH value
   c. dissociates close to 100%
   d. reacts with bases

5. The species in a phosphate buffer at pH 6.5 that reacts with added hydroxide ion is:
   a. $H_2PO_4^-$
   b. H_3PO_4
   c. HPO_4^{-2}
   d. PO_4^{-3}

6. The formula weight of a triprotic acid is 144. The equivalent weight is:
   a. 432
   b. 72
   c. 48
   d. 288

7. The Ka value for lactic acid is 1.4 x 10^{-4}. The pKa value is:
   a. 4.1
   b. 3.8
   c. 4.6
   d. 3.1

8. A solution with pH = 7.7 has a [H$^+$] of:
   a. 2 x 10^{-7} M
   b. 7 x 10^{-7} M
   c. 2 x 10^{-8} M
   d. 7 x 10^{-8} M

9. The pKa values for four acids are 2.6, 3.6, 4.9 and 5.2. The weakest acid is the one with pKa value of:
   a. 2.6
   b. 3.6
   c. 4.9
   d. 5.2

10. An aqueous solution made up of 0.1 M sodium acetate (NaCH$_3$CO$_2$) will be:
    a. acidic
    b. basic
    c. neutral
    d. cannot determine

11. The pH of a solution of pH 8.0 does not change upon addition of strong acid. The original solution:
    a. contains a strong acid
    b. contains an indicator
    c. is a buffer solution
    d. tastes sour

12. The normality of 100 mL of 0.4 M H$_2$SO$_4$ solution is:
    a. 0.8
    c. 0.2

Chapter 8   Acids and Bases

b. 0.4   d. 0.04

13. The normality of a solution made up of 2×10^{-4} moles of $Mg(OH)_2$ in 400 mL of water is:
    a. 1×10^{-6} N
    b. 5×10^{-5} N
    c. 1×10^{-3} N
    d. 5×10^{-6} N

14. Considering the acids below, together with the corresponding conjugate bases, which buffer system would be used to produce a buffer solution at pH 9.0?
    a. acetic acid, $K_a = 1.8 \times 10^{-5}$
    b. boric acid, $K_a = 7.3 \times 10^{-10}$
    c. formic acid, $K_a = 1.8 \times 10^{-4}$
    d. carbonic acid, $K_a = 4.3 \times 10^{-7}$

15. The Henderson-Hasselbach equation is used to calculate which of the following:
    a. whether an acid is a strong or weak acid
    b. whether a reaction between an acid and base will go to completion
    c. the extent of acidosis or alkalosis in the blood
    d. the ratio of the concentrations of the acid and its conjugate base to be used in preparing a buffer solution at the desired pH.

16. A buffer may be prepared by mixing equal amounts of a
    a. weak acid and its conjugate base
    b. weak acid and NaOH
    c. strong acid and its conjugate base
    d. strong acid and a weak base

COMPLETION. Write the word, phrase or number in the blank space in answering the question.

1. Complete and balance the following reactions.
    a. HCl + NaOH → _____ + _____ + _____

    b. Na_2O + HNO_3 → _____ + _____ + _____

    c. H_2SO_4 + _____ → K_2SO_4 + $2H_2O$

    d. NH_3 + HCl → Cl^- + _____

    e. Na_2CO_3 + HI → _____ + _____ + _____
        final products

2. A 100 mL sample of unknown acid is titrated with 0.42 N NaOH. The end-point of the titration requires 29 mL of base. The normality of the acid is _____.

3. If the acid in question 2 were H_2SO_4, the molarity of the acid would be _____.

4. Indicate (yes or no) whether the following bases are Arrhenius and/or Bronsted-Lowry bases.

	Arrhenius	Bronsted-Lowry
a. NH_3		

Chapter 8   Acids and Bases   65

    b.   NaOH
    c.   Al(OH)$_3$

5. The term hydronium ion is used interchangeably with _____ , _____ and _____.

6. The _____ of an acid in an intrinsic property of an acid, while the concentration is a variable and depends on the preparation of the solution.

7. Indicate the conjugate acid or base to the species below.
    a.   H$_3$O$^+$ _____ (conjugate base)

    b.   H$_2$O   _____ (conjugate base)

    c.   HSO$_4^-$ _____ (conjugate acid)

    d.   Cl$^-$ _____ (conjugate acid)

8. Indicate whether the acids shown below are mono-, di- or triprotic acids.
    a.   H$_3$PO$_4$ _____

    b.   H$_3$CCOOH   _____

    c.   H$_2$S   _____

    d.   H$_2$CO$_3$ _____

    e.   H$_2$PO$_4^-$ _____

9. Complete the following table

	[H$^+$]	[OH]	pH
a.	1 x 10^{-2} M	_____	____
b.	_____	2.5 x 10^{-3} M	____
c.	_____	_____	6.4

10. Indicate whether the dissolution of the following salts will produce a neutral, acidic or basic solution.
    a.   KCl   _____

    b.   NH$_4$I   _____

    c.   Na (HCO$_2$) (formic acid   _____
            is a weak acid)

    d.   NaClO$_4$ _____

11. If the pH of the blood goes below the normal pH, the condition is called _____. This may be caused by _____, a difficulty in breathing.

Chapter 8 Acids and Bases

ANSWERS TO SELF-TEST QUESTIONS

Multiple Choice
1. c
2. a, c
3. b, c
4. c, d
5. a
6. c
7. b
8. c
9. d
10. b
11. c
12. a (note that the volume has no influence on normality)
13. c
14. b
15. d
16. a

Completion
1. a. H_2O, Na^+, Cl^-
   b. H_2O, 2 Na^+, 2 NO_3^-
   c. 2 KOH
   d. NH_4^+
   e. H_2O, $CO_2\uparrow$, 2 NaI
2. 1.4 N
3. 0.7 M
4. a. no, yes
   b. yes, yes
   c. yes, yes
5. proton, H^+, H_3O^+
6. strength
7. a. H_2O
   b. OH^-
   c. H_2SO_4
   d. HCl
8. a. triprotic
   b. monoprotic
   c. diprotic
   d. diprotic
   e. diprotic
9. a. 1×10^{-12} M, 2
   b. 4.0×10^{-12}, 11.4
   c. 4×10^{-7}, 2.5×10^{-8}
10. a. neutral; salt of a strong acid (HCl) and a strong base (KOH)
    b. acidic, salt of a strong acid (HI) and a weak base (NH_4OH)
    c. basic; salt of a weak acid (HCO_2H) and a strong base (NaOH)
    d. neutral; salt of a strong acid ($HClO_4$) and a strong base (NaOH)
11. acidosis, hypoventilation

Every man, woman and child inhabiting this planet
became an unwitting guinea pig in a vast
biochemical experiment on July 16, 1945, when
the first atomic bomb was successfully detonated.
 - Ernest Borek
   The Atoms Within Us

Nuclear Chemistry

CHAPTER OBJECTIVES

After you have studied the chapter and worked the assigned exercises in the text and the study guide, you should be able to do the following.

1. Characterize the three major types of radioactivity in terms of symbol, charge, mass and penetrating capability.

2. Associate the names of famous scientists with their accomplishment in nuclear chemistry.

3. Given the frequency (v) of electromagnetic radiation and the equation $(\lambda)(v) = c$, calculate the wavelength (λ). Given the wavelength, calculate the frequency.

4. Order the different forms of radiation in the electromagnetic spectrum in terms of increasing energy and frequency and decreasing wavelength.

5. Explain the difference between a chemical and a nuclear reaction.

6. Write balanced nuclear equations for a reaction involving an (i) alpha particle, (ii) beta particle and (iii) gamma emission. Note whether the product nuclei has a mass greater than, equal to or less than the original radioactive nuclide.

7. Given an equation which includes all but one particle in a nuclear reaction, balance the equation and identify the missing particle.

8. Explain the term "half-life" and relate this to the decay curve for a radioactive nuclide.

9. Explain how a Geiger-Muller counter functions in monitoring the intensity of beta particle emission.

10. Define the three units, roentgens, rems and rads, used to describe the effects of radiation and indicate how they relate to each other.

11. Explain how ionizing radiation damages tissue by the formation of free radicals.

12. Write out the relationship which describes how the intensity of ionizing radiation varies with distance.

13. Describe the role of cobalt-60, iodine-131, and technetium-99m in medical diagnosis and/or treatment.

14. Explain the importance of fusion reactions to life on earth.

15. Point out the most obvious differences between nuclear fusion and fission reactions. Outline how the fission reactions can be used to produce controlled nuclear energy on the one hand and the atomic bomb on the other hand.

16. Define the important terms and comparisons in this chapter and give specific examples where appropriate.

IMPORTANT TERMS AND COMPARISONS

Alpha (α) Particle, Beta (β) Particle, and Gamma (γ) Ray
Wavelength (λ) and Frequency (υ)
$(\lambda)(\upsilon) = c$
Photon and electron
Electromagnetic Spectrum
Subatomic Particle
Nuclide
Mass Number and Atomic Number
Hydrogen, Deuterium and Tritium
Isotope
Alpha, Beta or Gamma Emitter
Natural and Artificial Transmutation
Half-Life ($T_{1/2}$)
Radioactive Decay Curve
"Phosphor"
Geiger-Muller Counter
Spectrum of Radiation
$(I_1)(d_1^2) = (I_2)(d_2^2)$

Roentgen, Rad and Rem
Equivalent Dose
Radiation (Exposure) Badge
Mutation
Background Radiation
Hyperthyroidism
Tagging a Drug
Collimated Radiation Beam
Radiation Sickness
"Big Bang" Explosion
Fusion
$E = mc^2$
Isotope Generator
Nuclear Fission and Fusion
Nuclear Power Plant
Control Rod
Nuclear Waste
Breeder Reactor
CAT, MRI, PET and BNCT

SELF-TEST QUESTIONS

MULTIPLE CHOICE. In the following exercises, select the correct answer from the choices listed. In some cases, more than one choice will be correct.

1. A 4 g sample of a radionuclide at room temperature has a half-life of three months. The amount of sample remaining after a year is:

Chapter 9 Nuclear Chemistry

    a.   0.25 g   c.   2 g
    b.   0.40 g   d.   4 g

2. If the radioactive decay of the sample in exercise 1 were carried out -100°C, the decay process would:
    a.   decrease   c.   remain unchanged
    b.   increase   d.   cannot determine

3. A nucleus which emits a beta particle will be transformed into a product nucleus of the:
    a.   same mass number
    b.   same atomic number
    c.   atomic number one unit greater
    d.   mass number one unit greater

4. The wavelength of blue light is 4000 A (1 A = 10^{-8} cm). Its frequency is:
    a. 12×10^{-12} s^{-1} c. 7.5×10^{10} s^{-1}
    b. 7.5×10^{14} s^{-1} d. 12×10^{10} s^{-1}

5. A scintillation counter uses a "phosphor" to:
    a.   determine $t_{1/2}$ values
    b.   produce transuranium elements
    c.   measure the number of mutations
    d.   measure the radioactivity in a sample

6. A millicurie is a unit of radioactivity equivalent to:
    a. 3.7×10^{10} counts s^{-1} c. 3.7×10^{7} counts s^{-1}
    b. 3.7×10^{10} counts min^{-1} d. 3.7×10^{4} counts s^{-1}

7. The radiation which causes the most damage within the cell is:
    a.   alpha particles   c.   gamma rays
    b.   beta particles   d.   X-rays

8. Which of the following radioactive nuclides decay the fastest?
    a. $^{197}_{80}$Hg, $t_{1/2}$ = 65 hrs   c. $^{238}_{92}$U, $t_{1/2}$ = 4×10^{9} yr.
    b. $^{210}_{84}$Po, $t_{1/2}$ = 138 days   d. $^{131}_{53}$I, $t_{1/2}$ = 8 days

9. The wavelength of which radiation is the greatest (longest)?
    a.   microwaves   c.   visible light
    b.   X-rays   d.   radiowaves

10. The number of neutrons in phosphorus-32, 32P is:
    a.   32   c.   17
    b.   15   d.   5

11. Which of the following nuclear reactions does not involve natural transmutation?
    a. $^{11}_{5}$B → $^{11}_{5}$B + γ

Chapter 9 Nuclear Chemistry

b. $^{210}_{84}Po \rightarrow\ ^{206}_{82}Pb +\ ^{4}_{2}He$

c. $^{14}_{6}C \rightarrow\ ^{14}_{7}N +\ ^{0}_{-1}e$

12. A source of ionizing radiation is found to have an intensity of 25 mCi at a distance of 1 m. To reduce the intensity of this radiation to 1 mCi, calculate what distance would be needed.
    a. 5 m
    b. 25 m
    c. 2 m
    d. 10 m

COMPLETION. Write the word, phrase or number in the blank space in answering the question.

1. Consider the reactions in question 11 in the multiple choice section. Reaction a) is an example of a(n) _____ emitter, reaction b) shows a(n) _____ emitter, while reaction c) exhibits a(n) _____ emitter.

2. Complete the following nuclear reactions. Refer to a periodic table.
    a. $^{28}_{13}Al \rightarrow \beta +$ _____

    b. _____ $\rightarrow\ ^{1}_{1}H +\ ^{0}_{-1}e$

    c. $^{254}_{102}No \rightarrow\ ^{4}_{2}He +$ _____

    d. $^{87}_{36}Kr \rightarrow\ ^{87}_{37}Rb +$ _____

3. The order of increasing penetrating power for gamma rays, alpha and beta particles is:

   _____ < _____ < _____

4. Match up the names of the famous scientist in the left-hand column with accomplishments or subjects most associated with the individual.
    a. H. Becquerel   (i)   Discovered X-ray
    b. E. Rutherford   (ii) $E = mc^2$
    c. G. Seaborg   (iii) CAT scan
    d. A. Einstein   (iv) Prepared transuranium elements
    e. W. Roentgen   (v)   "Coined" the names alpha, beta and gamma rays
    f. A. Cormack and G. Hounsfield (vi) 3.7×10^{10} counts s^{-1}
    g. M. Curie   (vii) Observed that certain rocks gave off mysterious radiation

5. Radiation that is derived from both cosmic rays and rocks in our environment contribute to the so-called _____ radiation.

6. Another name commonly used for an alpha particle is the _____ of a

Chapter 9 Nuclear Chemistry   71

_____ atom.

7. A beta particle has mass number of _____ and an atomic number of _____.

8. Cosmic rays are streams of _____, while gamma rays consist of _____, _____.

9. Isotopes have the same number of _____ but different number of _____.

10. A sample of a radioactive nuclide with a half-life of 10 years is left in a cave for 35 years. After this time, only 0.5 grams remains. The amount of this radioactive material originally left in the cave was _____.

11. A sample of radioactive isotope has an activity of 4.0 mCi. This sample emits _____ counts per second.

12. Diagnostic X-rays are absorbed by the bones and not the _____ because the bones contain elements of _____ atomic number which absorb X-rays strongly.

13. Consider the following two reactions:
   a. $^{252}_{98}Cf + ^{10}_{5}B \rightarrow ^{257}_{103}Lr + 5\,^{1}_{0}n$

   b. $^{235}_{92}U + ^{1}_{0}n \rightarrow ^{90}_{37}Rb + ^{144}_{55}Cs + 2\,^{1}_{0}n$

   Reaction (a) is an example of a _____ reaction, while reaction (b) is a _____ reaction.

14. An important characteristic of a breeder reaction is that it not only provides _____, but also makes additional _____.

15. A _____ _____ is a compound with _____ electrons and as a result is highly reactive.

16. Medical diagnosis has recently made significant advances in that physicians can diagnose many medical problems by non-invasive procedures (no surgical procedures). Complete the following table which characterizes three of these procedures.

Acronym	Complete Name	Probe (Radiation) Used In Diagnosis
CAT (scan)		
PET (scan)		
MRI		

                    ANSWERS TO SELF-TEST QUESTIONS
Multiple Choice
 1. a   7. a
 2. c   8. a
 3. a, c   9. d
 4. b   10. c
 5. d   11. a

Chapter 9 Nuclear Chemistry

6. c   12. a

Completion

1. a. gamma
   b. alpha
   c. beta
2. a. $^{28}_{14}Si$
   b. $^{1}_{0}n$
   c. $^{250}_{100}Fm$
   d. $^{0}_{-1}e$
3. alpha < beta < gamma
4. a. (vii)   d. (ii)
   b. (v)   e. (i)
   c. (iv)   f. (iii)
                  g. (vi)
5. background
6. nucleus, helium
7. zero, minus one
8. nuclei (mainly protons), high energy radiation
9. protons, neutrons
10. ca. 6 g
11. 1.5×10^8
12. soft tissue, higher
13. fusion, fission
14. energy, fuel
15. free radical; unpaired
16. | | | |
    |---|---|---|
    | CAT (scan) | Computer Assisted Tomography | X-rays |
    | PET (scan) | Positron Emission Tomography | Isotopes emitting positrons |
    | MRI | Magnetic Resonance Imaging | Radio-frequency waves |

Chapter 10   Organic Chemistry

Gasoline - 29¢/gal
   - Sign at gas station in
     Forty Fort, PA; circa 1965

10

Organic Chemistry

CHAPTER OBJECTIVES

After you have studied the chapter and worked the assigned exercises in the text and the study guide, you should be able to do the following.

1. Indicate the importance of the experiment that F. Wöhler carried out in which he produced urea from ammonium chloride and silver cyanate.

2. Review the octet rule and how to draw Lewis structures for simple organic molecules, as outlined in Chapter 3.

3. Review the valence-shell electron-pair repulsion (VSEPR) model in Chapter 3 and predict the bond angles in simple organic molecules.

4. Recognize and structurally define the functional group of these types of compounds: alcohols, aldehydes, ketones, carboxylic acids, and amines.

5. Explain the difference in structural formula between a primary, secondary and tertiary alcohol and draw a formula for several of each of them.

6. Explain the difference in structural formula between a primary, secondary, and tertiary amine and draw a formula for several of each of them.

7. Recognize the difference between an aldehyde, ketone, and carboxylic acid functional groups, all of which contain a carbonyl group, and be able to draw each of them

8. Define the important terms and comparisons in this chapter and give specific examples where appropriate.

IMPORTANT TERMS AND COMPARISONS

Alcohols
Aldehydes

Chapter 10 Organic Chemistry

Amines
Carboxylic Acids
Functional Groups
Ketones
Lewis Structures
Molecular and Structural Formula
Organic Chemistry
Primary, Secondary and Tertiary Alcohols
Primary, Secondary and Tertiary Amines
Synthetic and Natural Compounds
VSEPR Model

SELF-TEST QUESTIONS

MULTIPLE CHOICE. In the following exercises, select the correct answer from the choices listed. In some cases, more than one choice will be correct.

1. The molecular formula, C_3H_8O, may represent the following class(es) of organic compound(s). Indicate which one is correct.
    a. aldehyde
    b. carboxylic acid
    c. amine
    d. alcohol

2. Indicate which of the following Lewis structures is correct.

    (a)  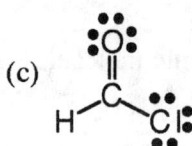 (b) 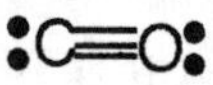 :C≡O:

    (c) 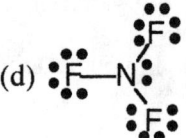 (d) :F—N:F: :F:

3. Indicate what type of amine the molecule, ethylamine, $CH_3CH_2NH_2$, is
    a. primary amine
    b. secondary amine
    c. tertiary amine
    d. quaternary amine

4. A tertiary carbon is bonded directly to:
    a. 2 hydrogens
    b. 3 carbons
    c. 2 carbons
    d. 4 carbons

5. The angle between the carbon atoms in propane, $CH_3CH_2CH_3$, is
    a. 180°
    b. 109.5°
    c. 120°
    d. 100°

COMPLETION. Write the word, phrase or number in the blank space or draw the appropriate structure in answering the question.

1. An aldehyde contains a _____ group, with the requirement that the carbonyl carbon atom must be the _____ carbon atom in the molecule.

Chapter 10 Organic Chemistry

2. A ketone contains a _____ group and it must be positioned at an _____ position in the molecule.

3. Cholesterol is made in the liver (natural) and also has been made synthetically in the laboratory. A comparison of the structure and its chemical characteristics shows that the cholesterol made by these entirely different routes is _____.

4. Draw the condensed structural formula for a primary and a tertiary amine that has three carbons and the necessary number of hydrogens.

5. Predict these bond angles in the ketone, acetone.

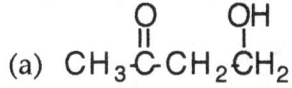

   a.   H-C-H _____   b.   H-C-C _____
   c.   O-C-C _____

6. Draw the Lewis structures for the following molecules.
   a. HCl (hydrogen chloride)
   b. HCOOH (formic acid)
   c. CH_3CH_2CHO (propanal)
   d. $H_2NCH_2CH_3$ (ethylamine)

7. Identify the functional groups in the following molecules.

   (a) $CH_3\text{-}\overset{\overset{O}{\|}}{C}\text{-}CH_2\text{-}\overset{\overset{OH}{|}}{C}H_2$

   (b) $CH_3\text{-}CH_2\text{-}\overset{\overset{O}{\|}}{C}\text{-}OH$

   (c) $CH_3\text{-}NH\text{-}CH_2\text{-}\overset{\overset{O}{\|}}{C}\text{-}CH_2\text{-}\overset{\overset{O}{\|}}{C}\text{-}H$

8. Draw structural formulas for at least four amines with the molecular formula of C_4NH_{11}.

9. Draw the condensed structure for a compound with the molecular formula indicated.
   a. a secondary alcohol with formula, C_3H_8O.
   b. a tertiary amine with formula, $C_4H_{11}N$.
   c. a molecule that has two aldehyde groups (a dialdehyde) with the formula, $C_4H_6O_2$.

ANSWERS TO SELF-TEST QUESTIONS
Multiple Choice
1. d
2. b, d
3. a
4. b
5. b. It is important to understand that in simple drawings, the carbons appear in a line (180°); however, the C-C-C) angles are 109.5°.

Completion
1. carbonyl, terminal or end
2. carbonyl, internal

3. identical
4. primary amine   tertiary amine
   CH$_3$CH$_2$CH$_2$NH$_2$ CH$_3$NCH$_3$
                                               |
                                              CH$_3$

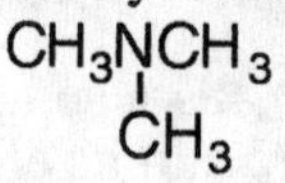

5. a. 109.5°
   b. 109.5°
   c. 120°

6. Remember that Lewis structures must show all valence electrons

   (a)  H-Cl: (b) 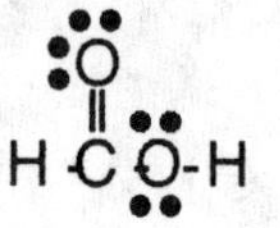 H-C(=O)-O-H

   (c) H-C(H)(H)-C(H)(H)-C(=O)-H   (d) 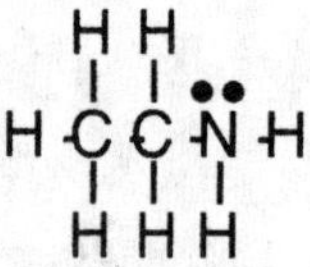 H-C(H)(H)-C(H)(H)-N(H)-H

7. a. a ketone and primary alcohol
   b. carboxylic acid
   c. an aldehyde, a ketone, and a secondary amine

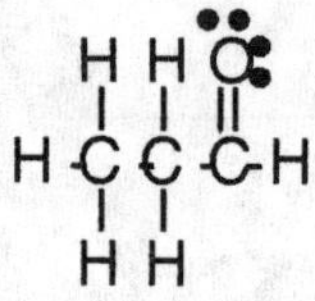

8. There are eight amines of this molecular formula.

   primary amines   CH$_3$CH$_2$CH$_2$CH$_2$NH$_2$ CH$_3$CH$_2$CH(NH$_2$)CH$_3$

                       CH$_3$CH(CH$_3$)CH$_2$NH$_2$ CH$_3$C(CH$_3$)(CH$_3$)NH$_2$

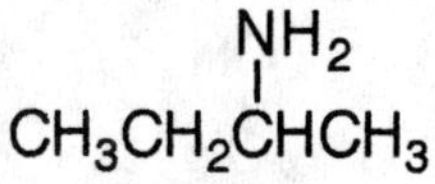

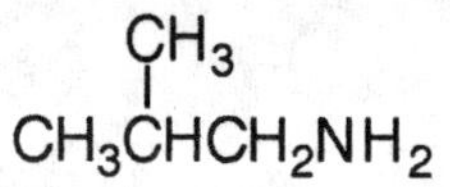

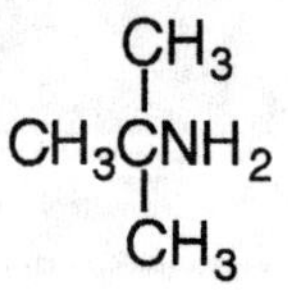

   secondary amines   CH$_3$CH$_2$CH$_2$NHCH$_3$ CH$_3$CH(CH$_3$)NHCH$_3$

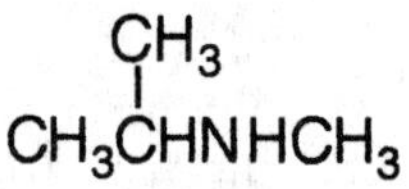

   tertiary amine   CH$_3$CH$_2$N(CH$_3$)CH$_3$

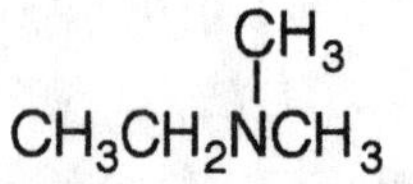

9. Following are structural formulas for each compound.

   (a) 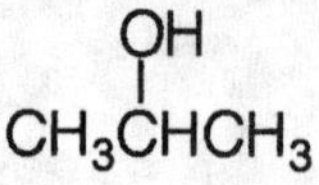 CH$_3$CH(OH)CH$_3$ (b) (CH$_3$)$_2$NCH$_2$CH$_3$ (c) 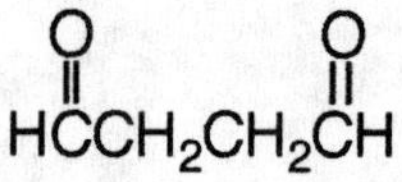 HC(=O)CH$_2$CH$_2$C(=O)H

Chapter 11 Alkanes and Cycloalkanes

Gasoline - 29¢/gal
- Sign at gas station in
   Forty Fort, PA; circa 1965

11

Alkanes and Cycloalkanes

CHAPTER OBJECTIVES

After you have studied the chapter and worked the assigned exercises in the text and the study guide, you should be able to do the following.

1. Name and draw structural formulas for the first ten unbranched alkanes, noting trends in their physical characteristics.

2. Consider the molecular formula for some alkanes with greater than 4 carbons and identify any constitutional or structural isomers.

3. Using the IUPAC rules of nomenclature for saturated hydrocarbons, write the names for a variety of unbranched alkanes, branched alkanes, and cycloalkanes.

4. Write the structural formulas of molecules for which the IUPAC names are given.

5. Memorize the prefixes, infixes, and suffixes that are used in naming molecules and provide information about the carbon-carbon bond and the functional group.

6. Become proficient in drawing the conformations for cyclic hydrocarbons and in distinguishing between axial and equatorial positions in cyclohexanes.

7. Recognize molecules that can exhibit cis and trans isomers.

8. Characterize the most important chemical and physical properties of alkanes.

9. Define the important terms and comparisons in this chapter and give specific examples where appropriate.

IMPORTANT TERMS AND COMPARISONS

Axial and Equatorial Bonds
Chair and Envelope Conformation

Chapter 11 Alkanes and Cycloalkanes

Cis and Trans Isomers
Combustion or Oxidation of Alkanes
Constitutional Isomers
Di-, Tri-, Tetra-, Penta- and Hexa-
Free Rotation and Conformations
Infixes -an-, -en- and -yl-
Iso, Sec- and Tert-
IUPAC Systematic and Common Names of Compounds
Lewis Structures
London Dispersion Forces
Molecular and Structural Formulas
Octane Rating
Unbranched and Branched Chains

SELF-TEST QUESTIONS
<u>MULTIPLE CHOICE</u>. In the following exercises, select the correct answer from the choices listed. In some cases, more than one choice will be correct.

1. The molecular formula, C_3H_8O, represents which organic functional group.
   a. alkane
   b. aldehyde
   d. amine
   e. alcohol

2. The family of hydrocarbons with a C=C (double) bond is called:
   a. alkene
   b. alkyne
   c. aldehydes
   d. amine

3. Which hydrocarbon has one CH_2 unit more than in $CH_3CH_2CH_2CH_2CH_2CH_2CH_3$?
   a. butane
   b. octane
   c. propylene
   d. heptane

4. Indicate which of the following is not a constitutional isomer for C_5H_{12}.

   a. $CH_3CH_2CH_2CH_2CH_3$

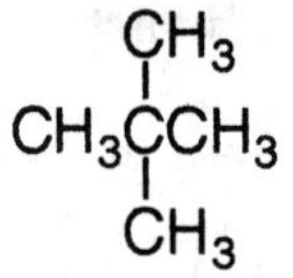

   b. 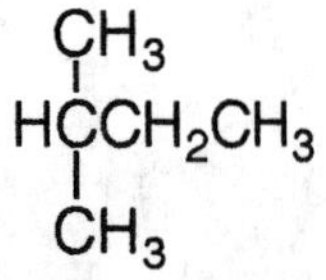 $CH_3\underset{CH_3}{\overset{CH_3}{\underset{|}{\overset{|}{C}}H}}CH_2CH_3$

   c. $CH_3\underset{CH_3}{\overset{CH_3}{\underset{|}{\overset{|}{C}}}}CH_3$

   d. 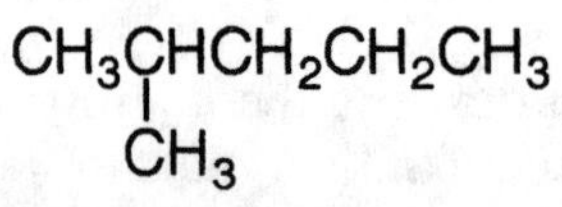 $CH_3\underset{CH_3}{\overset{|}{C}H}CH_2CH_2CH_3$

5. Which of the following structural formulas represent the same compound?

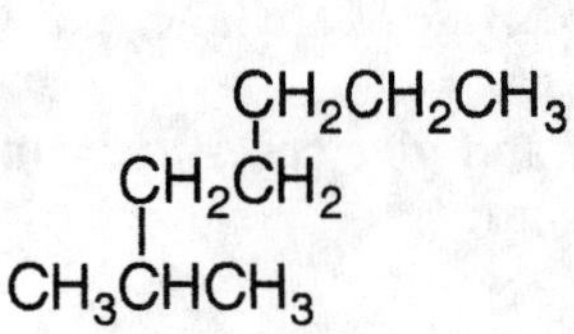

   (i) 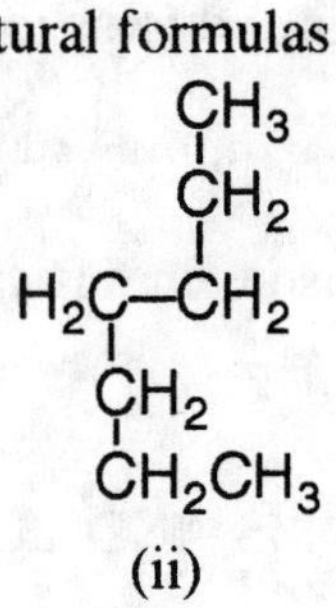   (ii) 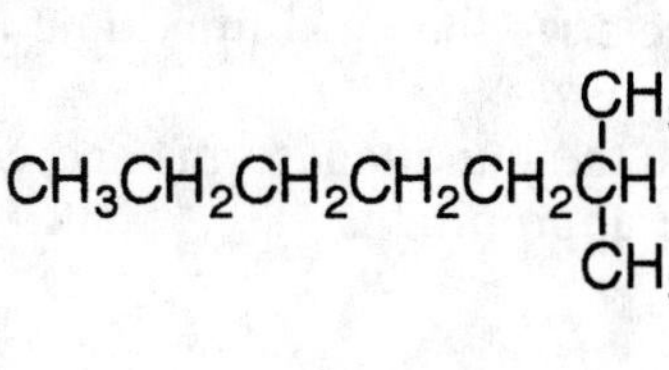   (iii)

a. all are different
b. (i) and (ii)
c. (i) and (iii)
d. all structural formulas

6. The IUPAC name for this compound is:

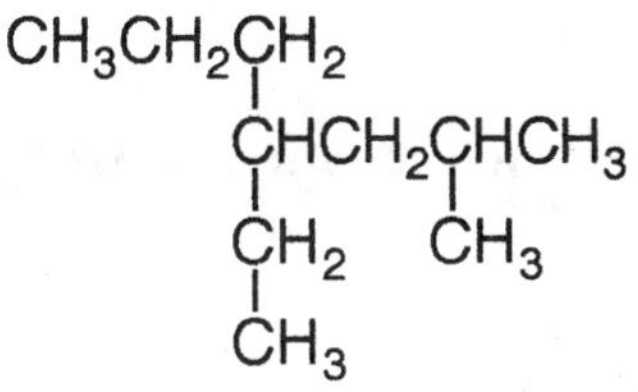

   CH₃CH₂CH₂
         |
         CHCH₂CHCH₃
         | |
         CH₂   CH₃
         |
         CH₃

   a. 2-methyl-4-propylhexane
   b. 4-ethyl-7-methylheptane
   c. 4-ethyl-2-methylheptane
   d. decane

7. The structure for 5-bromo-2-chloroheptane is:

   a. 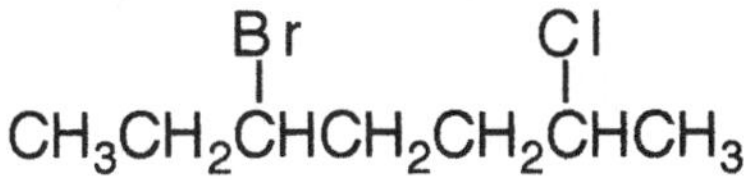 CH₃CH₂CHCH₂CH₂CHCH₃ (Br on C3, Cl on C6)

   b. 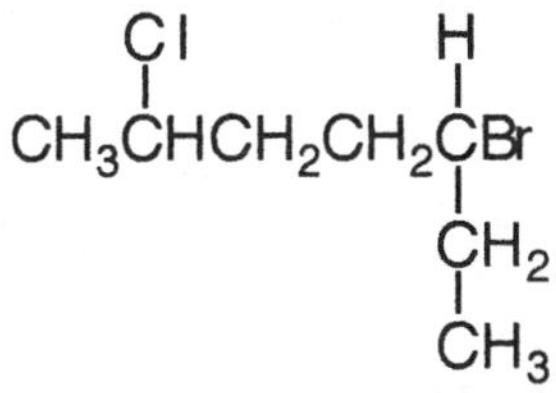 CH₃CHCH₂CH₂CBr (Cl on first C, H and CH₂CH₃ branch on last C)

   c. Br—⬡—Cl

   d. 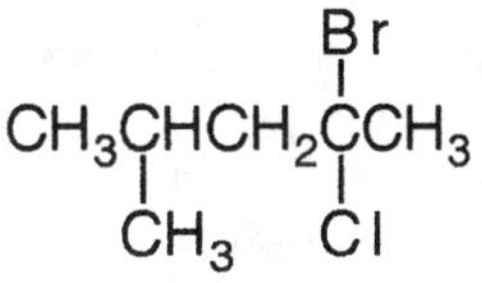 CH₃CHCH₂CCH₃ (Br and Cl substituents with CH₃)

8. The combustion of pentane produces:
   a. an alcohol
   b. carbon dioxide and water
   c. pentene
   d. ethane

9. Indicate which of the following sets of molecules are constitutional isomers.
   a. CH₃CH₂OCH₃ and CH₃CH₂CH₂OH
   b. cyclohexane and CH₂=CHCH₂CH₂CH₂CH₃
   c. CH₃OH and CH₃CH₂OH
   d. CH₃CH₂CHO and CH₃CH₂CH₂OH

10. Which is the IUPAC name of this compound?

    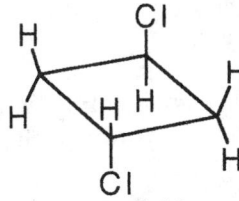

    a. trans-1,3-dichlorocyclobutane
    b. cis-1,2-dichlorocyclobutane
    c. cis-1,3-dichlorocyclobutane
    d. cyclobutane

11. Which hydrocarbon molecule with eight carbons would produce the lowest octane number (i.e., the most knocking) when used as a fuel in an automobile?
    a. 2,3,3-trimethylpentane
    c. 2-isopropylpentane

Chapter 11 Alkanes and Cycloalkanes

  b. octane   d. 2-methylheptane

COMPLETION. Write the word, phrase or number in the blank space or draw the appropriate structure in answering the question.

1. Examine the following structural formulas and supply the information required by writing the identifying number(s) in the blank.

  (i) $CH_3CH_2CH_2CH_2CH_3$

  (ii) $CH_3OCH_2CH_3$

  (iii) (structure: Cl, Cl on C=C with H and CH_2CH_3)

  (iv) (γ-butyrolactone-type ring with CH_2CH_3 substituent)

  (v) $H_3C-\underset{\underset{CH_2Cl}{|}}{\overset{\overset{CH_3CH_2}{|}}{C}}-CH_3$

  a. Saturated hydrocarbon   _____
  b. Contains aldehyde functional group _____
  c. Halogenated hydrocarbon   _____
  d. Unsaturated hydrocarbon   _____
  e. Unbranched hydrocarbon   _____
  f. Contains more than one functional group _____
  g. Acyclic hydrocarbon   _____

2. The outstanding chemical characteristic of alkanes is their _____.

3. Isomers are defined as molecules which have the same _____, but different _____.

4. The _____ positions in the chair conformation of cyclohexane are found pointing up and down, respectively, on adjacent carbons.

5. A cis isomer for 1,2-dimethylcyclopentane must the two methyl groups directed in the _____ direction on two _____ carbon atoms

6. The boiling point of a series of alkanes increases as the molecular weight of the alkane _____.

7. Cyclohexane in the chair form has (a number)_____ hydrogens "pointing" up or down called _____ hydrogens, in addition to the _____ hydrogens "pointing" approximately in the "plane" of the ring called _____ hydrogens.

8. Cis and trans isomers are examples of _____ or _____ isomers.

9. Write the IUPAC name for the each molecule:

Chapter 11 Alkanes and Cycloalkanes

a. 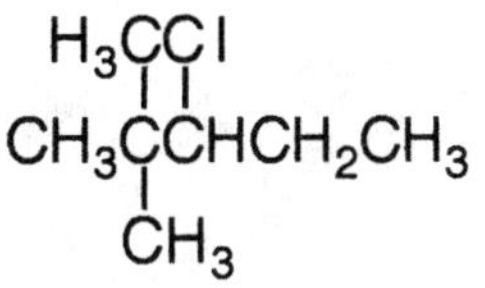 CH₃C(CH₃)(Cl)CH(CH₃)CH₂CH₃ _____

b. 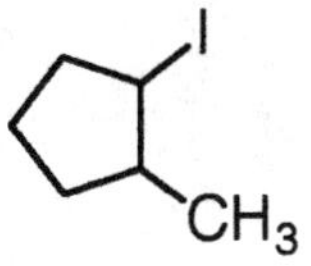 (cyclopentane with I and CH₃ substituents) _____

c. 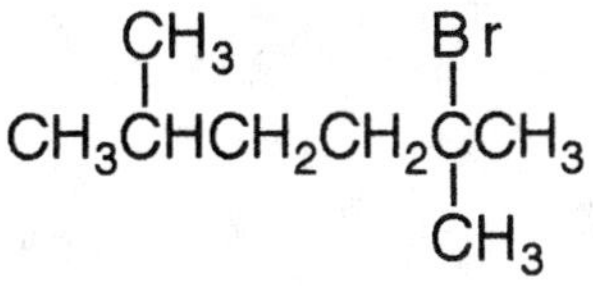 CH₃CH(CH₃)CH₂CH₂C(CH₃)(Br)CH₃ _____

10. Draw the structures for the following compounds, showing all the atoms.

   a. 3-chloro-2-methylpentane _____

   b. 1-iodo-3,3-dimethylcyclobutane _____

   c. 4-isopropylheptane _____

11. A saturated hydrocarbon can exist in a number of _____ because there is _____ rotation about a (C-C) single bond.

12. Balance the following equation for the combustion or oxidation of an alkane.

$$C_5H_{12} + O_2 \longrightarrow CO_2 + H_2O$$

ANSWERS TO SELF-TEST QUESTIONS

Multiple Choice
1. d
2. a
3. b
4. d
5. b
6. c
7. a and b (both are the same molecule)
8. b
9. a, b
10. a
11. b

Completion
1. a. i only; all other molecules contain other elements, in addition to C and H.
   b. none
   c. iii, v
   d. iii
   e. i
   f. iv; iii also, if each chlorine is considered as a functional group
   g. i, ii, iii, v
2. lack of reactivity
3. molecular formula, structural formula
4. axial
5. same, different

Chapter 11 Alkanes and Cycloalkanes

6. increases
7. 6, axial, 6, equatorial
8. cis-trans, stereoisomers
9. a. 3-chloro-2,2-dimethylpentane
   b. 1-iodo-2-methylcyclopentane
   c. 2-bromo-2,5-dimethylhexane
10. Structural formulas for each are:

    (a) 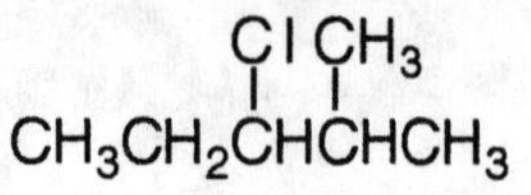 CH$_3$CH$_2$CH(Cl)CH(CH$_3$)CH$_3$

    (b) 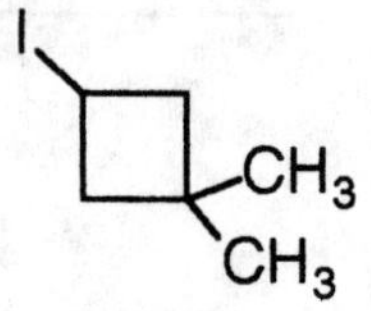 1-iodo-3,3-dimethylcyclobutane structure

    (c) 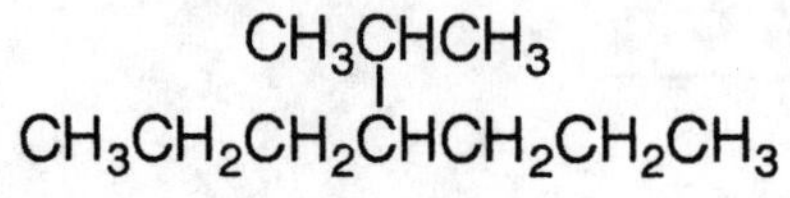 CH$_3$CH$_2$CH$_2$CH(CH(CH$_3$)CH$_3$)CH$_2$CH$_2$CH$_3$

11. conformations, free
12. the balanced equation is:

    $$C_5H_{12} + 8CO_2 \longrightarrow 5CO_2 + 6H_2O$$

Chapter 12   Alkenes and Alkynes

On Chemistry:
It's hard and its hard, ain't it
hard, Good Lord
   - Woody Guthrie
      "Hard, Ain't It Hard," 1952

12

Alkenes and Alkynes

CHAPTER OBJECTIVES

Chapters 11 and 12 begin to present some of the substance of organic chemistry. This includes not only the formal naming of, but also the drawing of structures for organic molecules containing a particular functional group. In addition, there will be an increasing emphasis on recognizing the physical properties, and especially the chemical reactions, that can take place as a direct consequence of the presence of the functional group (i.e., the "reactive center"). The strategies and learning patterns developed in studying the material in this chapter will be especially useful also in subsequent chapters.

After you have studied the chapter and worked the assigned exercises in the text and the study guide, you should be able to do the following specific objectives.

1. Recognize the differences in structural formulas between alkenes and alkynes, and distinguish them from aromatic (arene) compounds.

2. Name alkenes and alkynes using the IUPAC rules.

3. Given the IUPAC name for an alkene or alkyne, draw its structural formula.

4. Characterize the addition reactions in terms of the proposed intermediates and the products that occur as when an alkene is treated with H_2, HCl, and Cl_2.

5. Use Markovnikov's rule to predict the product formed by treating an unsymmetrical alkene with HX (X = Cl, Br, OH).

6. Outline the manner in which ethylene and substituted ethylenes undergo chain-growth polymerization to form useful polymers.

Chapter 12   Alkenes and Alkynes

7. Define the important terms and comparisons and give specific examples where appropriate.

IMPORTANT TERMS AND COMPARISONS

Alkane, Alkene, and Alkyne
IUPAC Nomenclature
Markovnikov's Rule
Monomer and Polymer
Reaction Types for Alkenes
    a. Addition
    b. Catalytic Hydrogenation
    c. Hydration
    d. Chain-Growth Polymerization
Regioselective Reaction
Unsaturated and Saturated Hydrocarbon

Before challenging your comprehension of this material by working out the SELF-TEST, a few words are in order to help sort out the major themes presented. As this and subsequent chapters are individually addressed, it will become clear that although the general objectives and the nature of the material are similar, the strategy in learning is much the same. This is especially true in Chapters 13 through 18. Each of these chapters introduces one or more families of molecules with a characteristic functional group. **IT IS THE FUNCTIONAL GROUP THAT IS THE REACTIVE CENTER OF THE MOLECULE.** That is, this is where the reaction will take place with other molecules. The functional group will be the dominant factor in defining the chemical and physical properties of a compound. Compare the chapter titles with the list of functional groups. The focus of Chapter 12 is aimed at basically three major aspects of the two families of unsaturated molecules. The three themes are:

1. The **naming** of unsaturated molecules using the internationally accepted IUPAC rules. This nomenclature systematically defines the composition and structure of the molecule.

2. The **characterization of the physical properties** of each family of molecules.

3. **Identification of the reactions** that characterize each type of unsaturated molecule. These same general themes will reoccur in the subsequent chapters, with simply the focus reset on a different functional group.

Possibly the most troublesome area for students is developing a strategy in identifying and characterizing the reactions associated with a family of molecules. A guide or "road map" for the process, which is both simple and useful, is the following.

1. Identify the functional group.
2. Tabulate the reactions associated with the functional group.
3. Characterize the reaction by its general name (addition reaction, oxidation, substitution, etc.) and then tabulate any additional specifics (catalyst or heat required, follows Markovnikov's rule, etc.).

It is very helpful if the student can write out a condensed summary of the reactions for each family of molecules. An example for alkenes follows.

Chapter 12 Alkenes and Alkynes

Condensed Reaction Summary for Alkenes

Reaction	**Name**	**Comment**
(1)	Polymerization	Requires appropriate catalyst; addition reaction
(2)	Catalytic hydrogenation	Addition reaction; pressure and metal catalyst
(3)	Combustion or oxidation of a hydrocarbon	CO_2 and H_2O are the common combustion products of all hydrocarbons
(4)	Bromination, chlorination	Addition reaction
(5)	Addition HX	Obeys Markovnikov's rule
(6)	Hydration	Obeys Markovnikov's rule

SELF-TEST QUESTIONS

MULTIPLE CHOICE. In the following exercises, select the correct answer from the choices listed. In some cases, two or more answers will be correct.

1. A terpene containing 20 carbons has the following characteristics:
   a. 4 isoprene units
   b. 4 double bonds
   c. the product of a plant
   d. contains alkyne units

2. The reaction of cycloheptene with HCl yields
   a. 1,2-dichlorocycloheptene
   b. 2-chloroheptane
   c. chlorocycloheptane
   d. 1-chlorocylcohexane

3. The carbocation intermediate formed in the reaction of HBr with 3-methyl-2-butene is:

Chapter 12   Alkenes and Alkynes

a. 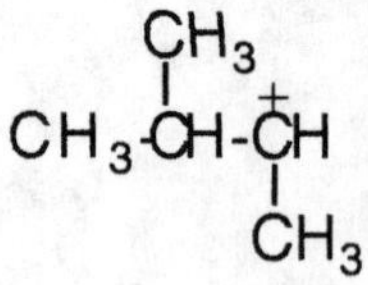 CH₃-CH-CH⁺
      | |
    CH₃ CH₃

c. 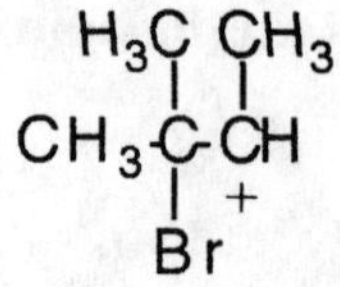 CH₃-C-CH⁺
    | |
  H₃C CH₃
       |
      Br

b. 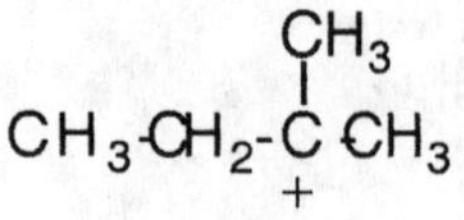 CH₃-CH₂-C⁺CH₃
          |
         CH₃

d. 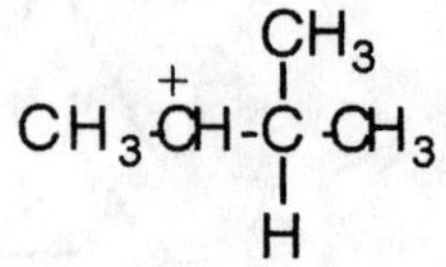 CH₃-CH-C⁺CH₃
       |
     CH₃
       |
       H

4. The principal product of the reaction of Br₂ with methylcyclopentene is:
   a. 2-bromo-1-methylcyclopentane
   b. 1, 2-dibromo-1-methylpentane
   c. no reaction
   d. 1, 2-dibromo-1-methylcyclopentane

5. The product of the reaction of HBr with which of these starting materials yields 2-bromo-2-methylbutane.

a. 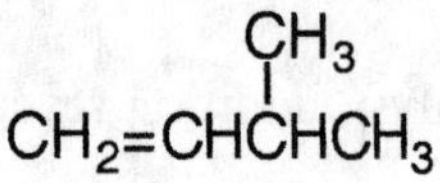 $CH_2=CHCHCH_3$
             |
          CH₃

c. 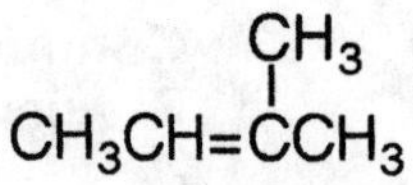 $CH_3CH=CCH_3$
             |
         CH₃

b. $CH_3CH_2CH_2CH_2CH_3$

d. 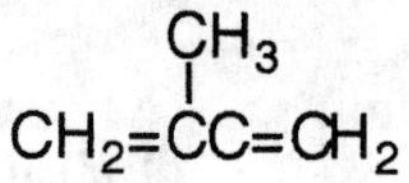 $CH_2=CC=CH_2$
        |
     CH₃

6. The polymerization of vinyl chloride ($H_2C=CHCl$) produces:
   a. unsaturated poly(vinyl chloride)
   b. saturated poly(vinyl chloride)
   c. a branched polymer
   d. a linear polymer

7. An example of a secondary carbocation is:

   a. CH_3^+

   c. 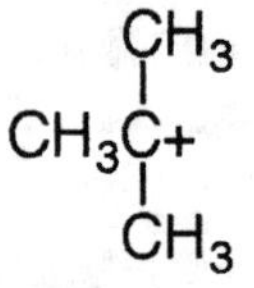 CH_3C^+
        |
     CH₃ (above and below)

   b. $CH_3CH_2^+$

   d. 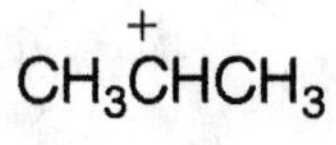 $CH_3\overset{+}{C}HCH_3$

8. The structural formula for the cis isomer of ClHC=CCH₃ is:

a. 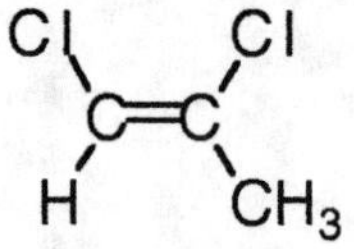 Cl\C=C/Cl with H and CH₃
b. 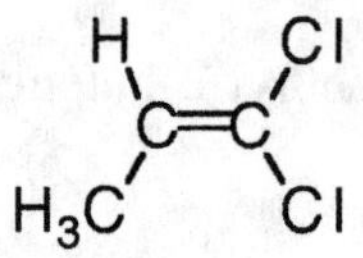 H\C=C/Cl with H₃C and Cl
c. Cl\C=C/CH₃ with H and Cl
d. 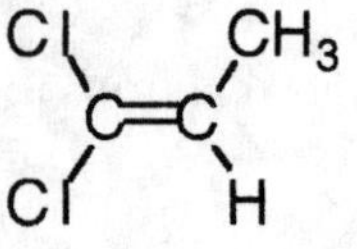 Cl\C=C/CH₃ with Cl and H

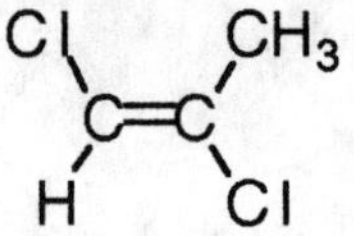

Chapter 12   Alkenes and Alkynes

9. The reaction of HBr with which of the compounds shown below, will produce CH₃CH₂CHBrCH₃:
   a. CH₂=CHCH₂CH₃
   b. CH₃CH=CHCH₃
   c. CH₃CH₂CH₂CH=CH₂
   d. CH₃C≡CCH₃

10. The IUPAC name for the compound below is:

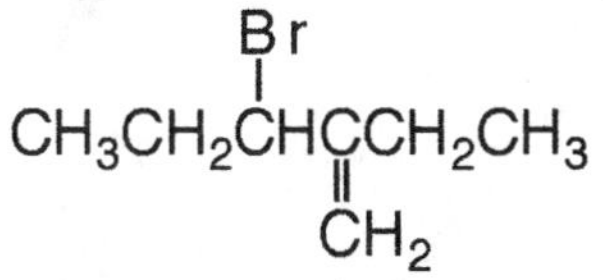

    CH₃CH₂CH(Br)C(=CH₂)CH₂CH₃

   a. 3-bromo-2-ethyl-1-pentene
   b. 3-bromo-2-ethyl-2-hexene
   c. 3-bromo-4-ethyl-1-pentene
   d. 3-bromo-4-ethyl-1-hexane

11. The name of the compound below is:

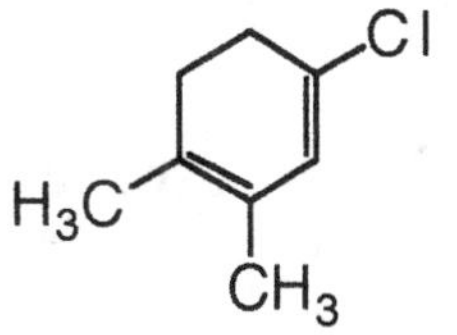

   a. 1-chloro-3,4-dimethyl-2,3-cyclohexatriene
   b. 4-chloro-1,2-dimethyl-1,3-cyclohexadiene
   c. 3,4-dimethyl-1-chloro-1,3-cyclohexadiene
   d. 1-chloroorthomethyl-1,4-cyclohexadiene

12. Indicate which of the carbocations shown below is the most stable and the least stable.
    a. CH₃CH₂CH₂CH₂⁺
    b. 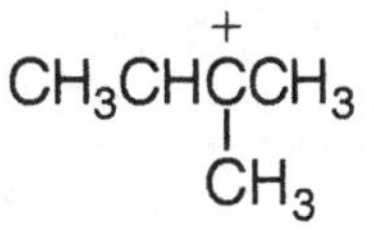 CH₃CH(CH₃)C⁺CH₃
    c. CH₃C⁺HCH₂CH₃
    d. 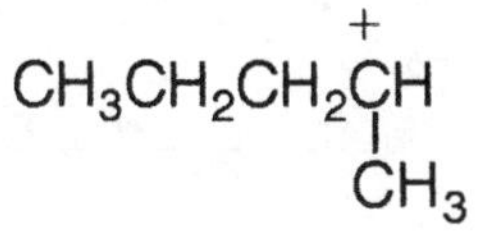 CH₃CH₂CH₂C⁺H(CH₃)

COMPLETION.   Write the word, phrase or number in the blank or draw the appropriate structure in answering the question.

1. Give the common or IUPAC names for the following molecules.
   a. 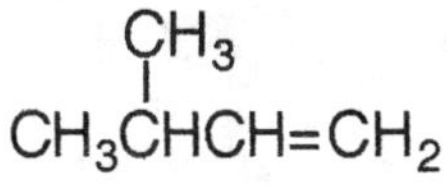 CH₃CH(CH₃)CH=CH₂
   b. ─(CH₂CH₂)ₙ─
   c.  (cyclopentane ring)
   d. (cyclohexadiene ring)
   e. CH₃CH₂(H)C=C(H)(H)C=C(CH₂CH₃)(H)

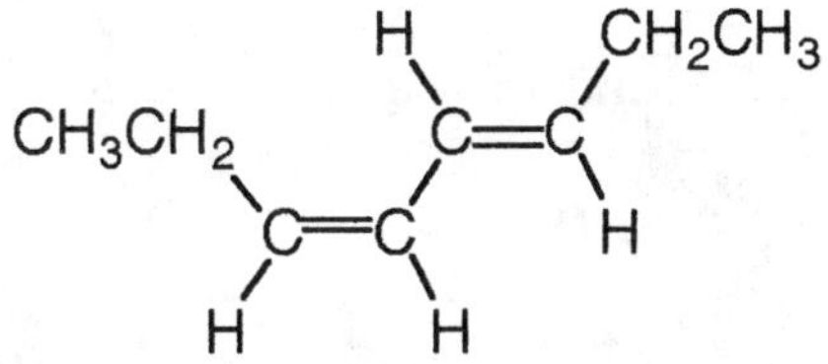

2. Fill in the Table below with the characteristics for alkanes, alkenes and alkynes.

Chapter 12 Alkenes and Alkynes

Characteristic	Alkanes	Alkenes	Alkynes
Infix			
Bonding Angle			

3. Draw the structures for the products of reactions b-d. Only indicate the products in reaction a.

   a. $CH_3CH=CH_2 + O_2 \longrightarrow$

   b. 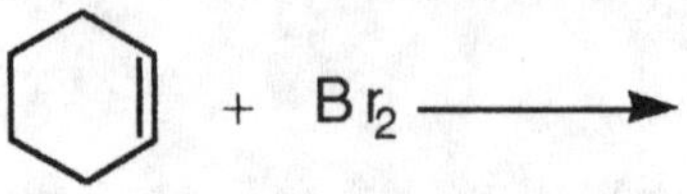 cyclohexene $+ Br_2 \longrightarrow$

   c. 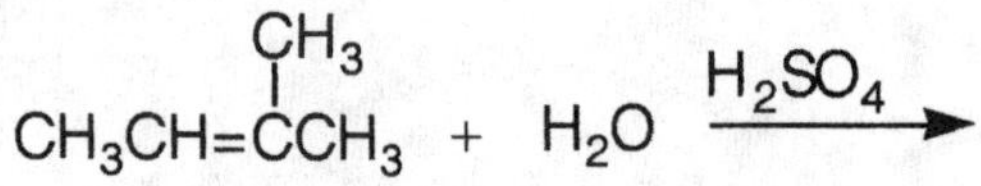 $CH_3CH=\overset{CH_3}{\underset{|}{C}}CH_3 + H_2O \xrightarrow{H_2SO_4}$

   d. 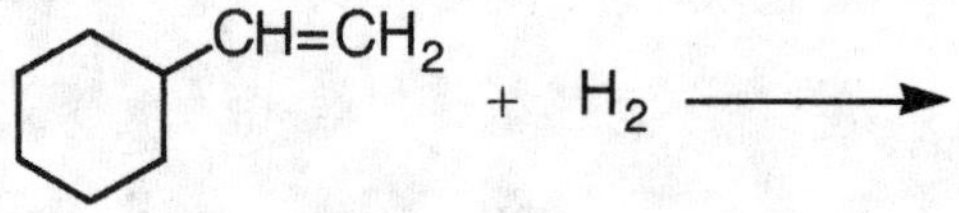 cyclohexyl-$CH=CH_2 + H_2 \longrightarrow$

4. Plastic combs are made of the polymeric material poly(methyl methacrylate) while indoor-outdoor carpeting is made of polypropylene. Draw the structural formulas for these products that are produced in the reactions below.

   a. 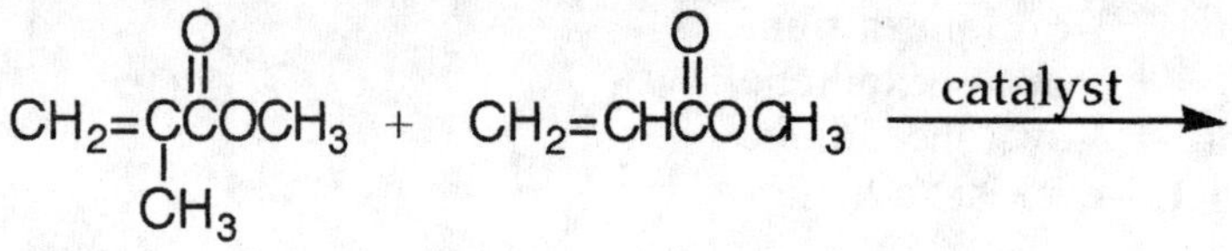 $CH_2=\overset{CH_3}{\underset{|}{C}}\overset{O}{\overset{\|}{C}}OCH_3 + CH_2=CH\overset{O}{\overset{\|}{C}}OCH_3 \xrightarrow{catalyst}$

   b. $CH_2=CHCH_3 \xrightarrow{catalyst}$

5. The boiling point for 1-hexene is _____ than that of 1-octene.

6. Markovnikov's rule states _____.

7. Draw the structural formula for 1, 2, 3-tribromo-4-chlorobutadiene.

8. Squalene is a terpene compound involved in the biosynthesis of cholesterol in the body. The structural formula is shown below and can be regarded as a polymer of isoprene monomer units.

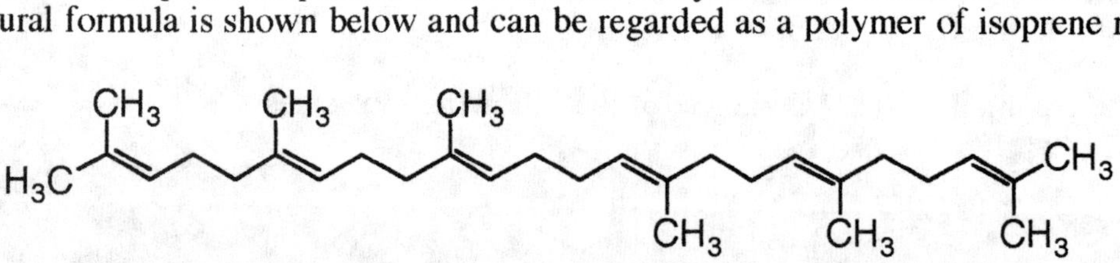

   [Structure of squalene shown]

   a. The structural formula for an isoprene unit is _____.

   b. Although the many structures of terpenes are very diverse, they all share the common characteristic _____.

   c. Determine the number of isoprene units in squalene and circle these units in the structure shown.

88

Chapter 12 Alkenes and Alkynes

ANSWERS TO SELF-TEST QUESTIONS

Multiple Choice
1. a, c
2. c
3. b
4. d
5. c
6. b, d
7. d
8. a
9. a, b
10. a
11. b
12. b, most stable; a, least stable

Completion
1.
   a. isoprene or 2-methyl-1, 3-butadiene
   b. polyethylene
   c. cyclopentene
   d. 1, 4-cyclohexadiene
   e. trans, cis-3, 5-octadiene

2.
Characteristic	Alkanes	Alkene	Alkynes
Infix	-an-	-en-	-yn-
Bonding Angle	109.5	120	180

3. Following are the products of each reaction.
   a. $CH_3CH=CH_2 + O_2 \longrightarrow CO_2 + H_2O$

   b. cyclohexene + $Br_2 \longrightarrow$ 1,2-dibromocyclohexane

   c. $CH_3CH=\underset{CH_3}{\overset{|}{C}}CH_3 + H_2O \xrightarrow{H_2SO_4} CH_3CH_2\underset{OH}{\overset{CH_3}{\underset{|}{\overset{|}{C}}}}CH_3$

   d. cyclohexyl-$CH=CH_2 + H_2 \longrightarrow$ cyclohexyl-CH_2CH_3

4. Answers to parts (a) and (b) show the repeting unit in each polymer.

   a. $+(CH_2-\underset{CH_3}{\overset{COOCH_3}{\underset{|}{\overset{|}{C}}}}-CH_2-\overset{COOCH_3}{\underset{|}{CH}})_n$

   b. $+(CH_2-\underset{|}{\overset{CH_3}{CH}})_n$

   c. addition
   d. two different monomer units
5. lower
6. When an unsymmetrical reagent such as HX adds to an alkene, the hydrogen adds to the carbon atom of the double bond which already has the greater number of hydrogen atoms.
7. This compound is:

Br-CH=C(Br)-C(Br)=CH-Cl

8. a. Isoprene is 2-methyl-1,3-butadiene

$$CH_2=C(CH_3)-CH=CH_2$$

b. All follow the isoprene rule.
c. 6 isoprene units

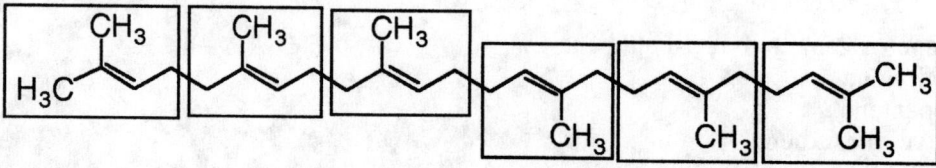

More wine, less wits, you know; wine makes a man sing even if he is a rare scholar, makes him titter and chuckle, aye, makes him dance a jig, and makes him blurt out what were better kept to himself.
   - Odyssey, XIV, 464

13

Alcohols, Ethers and Thiols

CHAPTER OBJECTIVES

After you have studied the chapter and worked the assigned exercises in the text and the study guide, you should be able to do the following.

1. Identify compounds that contain an alcohol, ether, or thiol functional group.

2. Given the IUPAC names for simple alcohols, draw a structural formula for each.

3. Given the structural formula for an alcohol, name the compound according to the IUPAC rules.

4. Write a reaction for the dehydration of an alcohol and point out the constraint put on the major product formed according to Zaitzev's rule.

5. Draw structural formulas for primary, secondary. and tertiary alcohols.

6. Write reactions for the oxidation of primary, secondary, and tertiary alcohols and characterize the possible products formed.

7. Review the concept of hydrogen bonding and discuss the role of hydrogen bonding in understanding the solubility characteristics of alcohols and ethers in water.

8. Characterize the similarities and differences in the physical properties of alcohols, ethers, thiols. and saturated hydrocarbons of comparable molecular weight.

9. Draw the structural formulas for thiols and disulfides and write an equation for the oxidation/reduction reaction associated with them.

Chapter 13 Alcohols, Ethers, and Thiols

10. Prepare a summary sheet for the reactions of alcohols, thiols and disulfides.

11. Define the important terms and comparisons in this chapter and give specific examples where appropriate.

IMPORTANT TERMS AND COMPARISONS
Alcohol and Sulfhydryl Groups
Alcohol, Ether and Thiol
Aldehyde and Ketones
Alfred Nobel
Anesthetic
Carboxylic Acid
Dehydration Reaction
Diol, Glycols and Triol
Elimination Reaction
Ethanol
Functional Group
Glycerol
Hydrogen Bonding
Hydrophobic
IUPAC Nomenclature
London Dispersion Forces
Nitroglycerine and Dynamite
Oxidation Reaction
Primary, Secondary, and Tertiary Alcohols
Thiols, Thioethers and Disulfides

SELF-TEST QUESTIONS

MULTIPLE CHOICE. Select the correct answer or answers from the choices listed. In some cases, more than one answer is correct.

1. Chromic acid oxidation of 2-butanol produces:

   a. $CH_3CH_2\overset{O}{\underset{\|}{C}}CH_3$ c. $CH_3CH_2\overset{O}{\underset{\|}{C}}H$

   b. $CH_3CH_2OCH_2CH_3$ d. $CO_2 + H_2O$

2. The compound that is most capable of hydrogen bonding is:
   a. $CH_3OCH_2CH_3$ c. $CH_3CH_2CH_2CH_3$
   b. $CH_3CH_2CH_2CH_2OH$ d. $CH_3CH_2SCH_2CH_3$

3. Indicate the type of chemical change that occurs in the alcohol during the breath test that law enforcement officers use to determine alcohol levels in drivers.
   a. oxidation   c. reduction
   b. dehydration   d. substitution

4. Indicate two characteristics of ethers that are dangerous.
   a. solubility in water
   b. flammability

Chapter 13 Alcohols, Ethers, and Thiols

    c.   potential for explosion
    d.   toxicity

5. The compounds which are widely recognized for their unpleasant odors are:
   a. thiols
   b. alcohols
   c. ethers
   d. alkanes

6. Tertiary alcohols cannot be oxidized because they:
   a. are already oxidized
   b. do not have a H on the C bearing the alcohol group
   c. are unsaturated
   d. readily form an alkene

7. The complete oxidation of a primary alcohol produces
   a. ketone
   b. aldehyde
   c. carboxylic acid
   d. cyclic ether

8. The oxidation of CH_3CH_2SH results in a
   a. thiol
   b. disulfide (bond formation)
   c. cyclic thiol
   d. useful anesthetic

9. The order of the compounds that increasingly take part in hydrogen bonding with water is
   a. alkane< alcohol<ether<aldehyde
   b. alcohol<alkane<aldehyde<ether
   c. alkane<ether<aldehyde<alcohol
   d. ether<aldehyde<alcohol<alkane

10. Oxidation of this alcohol by chromic acid yields:

    (structure: benzene ring with CH_2OH)

    a. benzene ring with CHO
    b. benzene ring with =O (cyclohexadienone)
    c. benzene ring with COOH
    d. benzene ring with CH_2CHO

11. The oxidation of ethanol produces acetaldehyde. Further oxidation of acetaldehyde yields
    a. HCOOH
    b. CH_3OH
    c. CH_3COOH
    d. CH_3COCH_3

12. Considering the potential for hydrogen bonding, predict what the relative boiling points for 1-propanethiol and ethylmethylsulfide would be.
    a. comparable
    b. ethylmethylsulfide is higher
    c. propanethiol is higher
    d. cannot predict

13. Zaitzev's rule must be invoked when which of the following alcohols are dehydrated in an acid-catalyzed reaction.
    a. methanol
    b. 2-pentanol
    c. 1-butanol
    d. ethanol

Chapter 13 Alcohols, Ethers, and Thiols   94

COMPLETION. Write the word, phrase or number in the blank or draw the appropriate structure in answering the question.

1. In the dehydration of an alcohol, two groups are removed from adjacent carbons to produce a double bond. This reaction often is referred to as an _____ reaction.

2. Tertiary alcohols cannot be _____ , while the oxidation of a primary alcohol yields either an _____ or a _____ .

3. The relative order of boiling points for a simple alcohol and an ether, both having comparable molecular weights is:

   _____ > _____

4. The relative solubility of 1-propanol, propane, 1,2-propanediol, and ethyl methyl ether in water is:

   _____ > _____ > _____ > _____

5. Complete the equations by writing the appropriate structural formula.

   a. 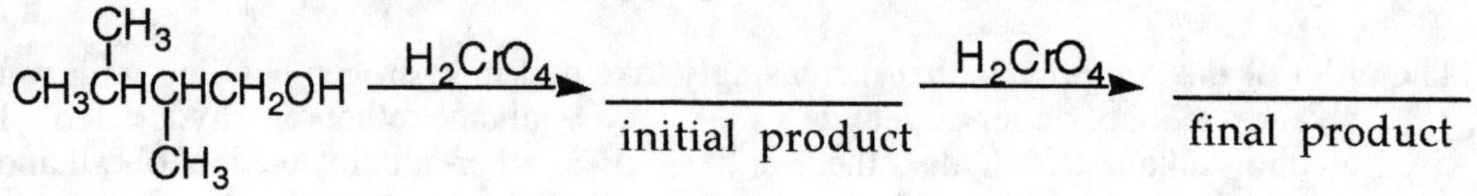 CH₃CHCHCH₂OH (with CH₃ groups on middle carbons) $\xrightarrow{H_2CrO_4}$ _____ initial product $\xrightarrow{H_2CrO_4}$ _____ final product

   b.  _____ $\xrightarrow{reduction}$ HSCH₂CHCH₂SH (with CH₃ on middle carbon)

   c. 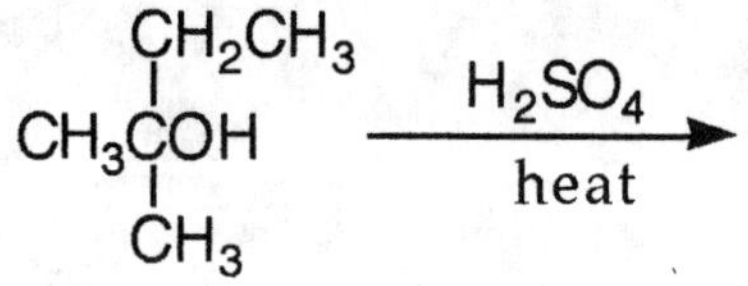 CH₃COH (with CH₂CH₃ and CH₃ groups) $\xrightarrow[heat]{H_2SO_4}$

   d. CH₃CH₂OH $\xrightarrow[heat]{H_2SO_4}$

   e. CH₃CH₂SH $\xrightarrow{oxidation}$

   f. 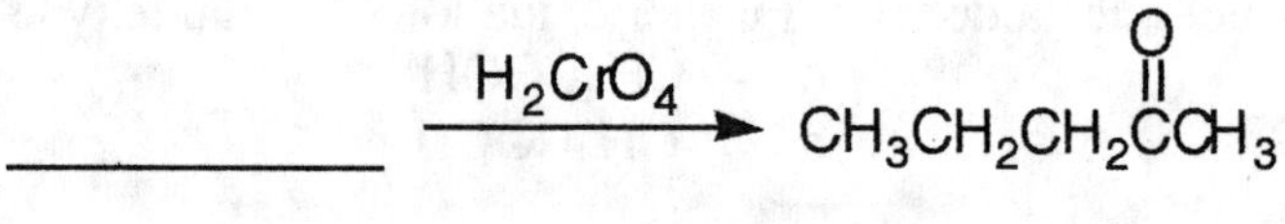 _____ $\xrightarrow{H_2CrO_4}$ CH₃CH₂CH₂CCH₃ (with C=O)

6. Characterize each of the following molecules as alcohols, ethers, both or none of these:

Chapter 13 Alcohols, Ethers, and Thiols   95

  a. CH$_3$CHO   _____

  b. CH$_3$CH$_2$CH$_2$OCH$_3$ _____

  c. HOCH$_2$CH$_2$OH   _____

  d. ⟨phenyl⟩–CH$_2$OH   _____

  e. ⟨cyclohexene with OH and CH$_2$OCH$_3$⟩ _____

7. Give suitable names for the following compounds.

  a. HOCH$_2$CHCH$_2$OH (with OH on middle carbon)

  b. CH$_3$CH$_2$OCH(CH$_3$)$_2$

  c. CH$_3$CHCCH$_3$ with HO and CH$_3$ on one carbon, CH$_3$ on the other

  d. ⟨cyclohexane with CH$_3$, OH, and CH$_3$CHCH$_3$ substituents⟩

  e. CH$_3$CH$_2$CH$_2$SH

8. Write out the structural formula for the following compounds:
  a. butyl methyl ether
  b. 4-ethyl-3-hepten-1-ol
  c. 1,2-pentanediol
  d. dichlorodifluoromethane
  e. the structure for any glycol.

9. In the reaction of potassium dichromate with ethanol, the dichromate is an _____ agent and becomes reduced in the reaction, while the alcohol is the _____ agent and becomes _____ in the reaction.

10. Explain how the drug, Antabuse (disulfiram), produces it effect in the treatment of alcoholism.

ANSWERS TO SELF-TEST QUESTIONS
Multiple Choice
1. a
2. b
3. a
4. b, c
5. a
6. b
7. c
8. b
9. c
10. a and/or c
11. c
12. a
13. b

Chapter 13 Alcohols, Ethers, and Thiols

Completion
1. elimination
2. oxidized; aldehyde, carboxylic acid
3. alcohol > ether
4. 1,2-propanediol > 1-propanol > ethyl methyl ether > propane
5.

    a. 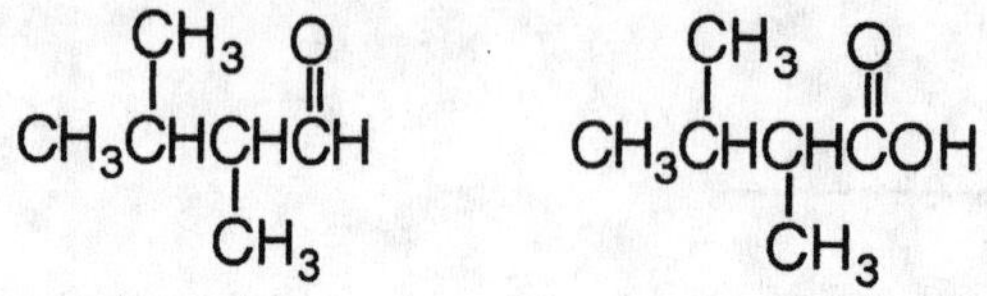 CH$_3$CH(CH$_3$)CH(CH$_3$)CHO   CH$_3$CH(CH$_3$)CH(CH$_3$)COOH   b. 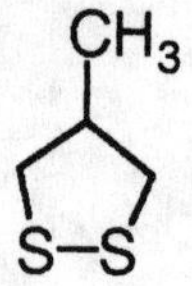 4-methyl-1,2-dithiolane ring

    c. 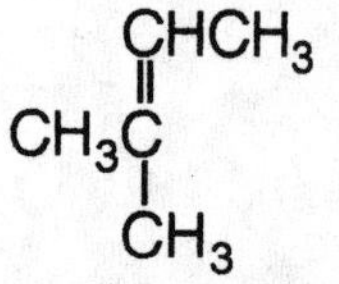 CH$_3$C(CH$_3$)=C(CH$_3$)CH$_3$ d. CH$_2$=CH$_2$

    e. CH$_3$CH$_2$S-SCH$_2$CH$_3$ f. 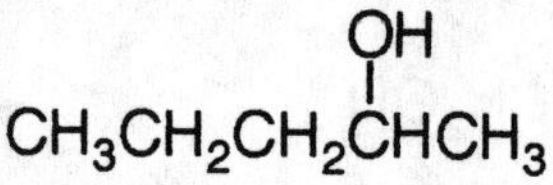 CH$_3$CH$_2$CH$_2$CH(OH)CH$_3$

6. a. none
   b. ether
   c. alcohol
   d. alcohol
   e. both an alcohol and an ether
7. a. 1, 2, 3-propanetriol
   b. ethyl isopropyl ether
   c. 3,3-dimethyl-2-butanol
   d. menthol or 2-isopropyl-5-methylcyclohexanol
   e. 1-propanethiol
8. Following are structural formulas:

    a. CH$_3$OCH$_2$CH$_2$CH$_2$CH$_3$ b. 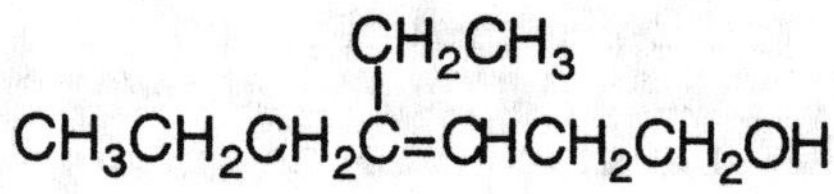 CH$_3$CH$_2$CH$_2$C(CH$_2$CH$_3$)=CHCH$_2$CH$_2$OH

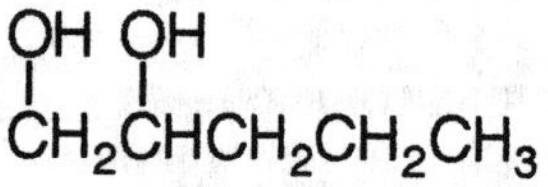

    c. CH$_2$(OH)CH(OH)CH$_2$CH$_2$CH$_3$ d. CCl$_2$CF$_2$

    e. any dialcohol with OH groups on adjacent carbons, for example
       HOCH$_2$CH$_2$OH

9. oxidizing, reducing, oxidized
10. Antabuse inhibits the (enzymatically catalyzed) reaction for the oxidation of acetaldehyde. The level of acetaldehyde builds up and produces nausea, sweating and vomiting.

A great teacher is one who makes
himself progressively unnecessary.
   - Thomas Carruthers

14

Benzene and Its Derivatives

CHAPTER OBJECTIVES

After you have studied the chapter and worked the assigned exercises in the text and the study guide, you should be able to do the following.

1. Recognize aromatic compounds and aromatic groups in larger, more general structures.

2. Draw a Lewis structure for the aromatic molecule, benzene, in its individual resonance forms.

3. Given the structural formula for an aromatic compound, write its the IUPAC name.

4. Given the IUPAC name for an aromatic compound, draw its structural formula.

5. Characterize the type of chemistry that aromatic molecules exhibit, including substitution reactions involving halogenation, nitration. and sulfonation.

6. Contrast the chemistry associated with alkenes with that for aromatic compounds.

7. Define the important terms and comparisons in this chapter and give specific examples where appropriate.

IMPORTANT TERMS AND COMPARISONS
Acid/Base Chemistry of Phenol
Addition Versus Substitution Reactions
Aromatic or Arene Compounds
Biodegradation
Carcinogen
DDT
Halogenation, Nitration and Sulfonation of Aromatic Compounds
K_a and pK_a Values and Acidity
Kekulé Structure
Ortho, Meta and Para

Chapter 14 Benzene and its Derivatives   98

Phenol and the Phenyl Group
Phenyl, and Aryl Groups
Polynuclear aromatic Hydrocarbon
Resonance and Aromatic Sextet

SELF-TEST QUESTIONS

MULTIPLE CHOICE. In the following exercises, select the correct answer from the choices listed. In some cases, two or more answers may be correct.

1. Indicate which of the following properties is a characteristic of benzene and its derivatives.
    a. Soluble in organic solvents and insoluble in water
    b. Burns in air to form carbon dioxide and water
    c. Bonding is essentially covalent
    d. Reacts with acids and bases

2. The nitration, halogenation and sulfonation of benzene are generally called
    a   substitution reactions   c.   elimination reactions
    b. addition reactions   d.   dehydration reactions

3. The name of the following molecule is

    a. 1, 2-dichloro-3, 4, 5-trimethylbenzene
    b. 5, 6-dichloro-2, 3-dimethyltoluene
    c. 2, 3-dichloro-4, 5, 6-trimethylbenzene
    d. 1, 2, 3-trimethyl-5, 6-dichlorobenzene

4. The individual (C-C) bonds in each Lewis structure for benzene are either single or double bonds. The (C-C) bond length in benzene is actually
    a, alternating single and double bonds
    b. a bond type midway between a single and double bond
    c. primarily single bonds
    d. primarily double bonds

5. In polynuclear aromatic hydrocarbons, the aromatic rings
    a. share a single carbon atom
    b. share two carbon atoms
    c. are held together by a single linking carbon atom
    d. in some cases, share more than one side

6. Reaction of iodine with the aromatic ring in the protein, thyroglobulin, starts the synthesis of thyroxine. Indicate what type of reaction this is.
    a. addition reaction   c.   substitution
    b. elimination   d.   oxidation

7. In a disubstituted benzene, the greatest possibility for the substituted groups "bumping into" each other is when they are in

Chapter 14   Benzene and its Derivatives   99

      a.   ortho positions   c.   meta positions
      b.   para positions   d.   all interactions are similar

8. Phenol exhibit a significant acidic character, much more than found in alcohols. The pK_a for phenol is 9.95. At pH 9.95, the relative ratio of phenol/phenoxide is
   a.   50/50   c.   100% phenol
   b.   30/70   d.   100% phenoxide

COMPLETION.   Write the word, phrase or number in the blank or draw the appropriate structure in answering the question.

1. Draw the structures for these reoccurring aromatic units.

    a.   phenol   _____

    b.   aniline   _____

    c   benzaldehyde   _____

    d.   benzoic acid   _____

    e.   anisole   _____

    f.   toluene   _____

2. Draw the structures for these reoccurring groups or substituents.

    a.   phenyl   _____

    b.   phenoxide   _____

3. Name the three different isomers that are possible for a benzene ring that has one nitro and one chloro substituent in place of hydrogens.

    a   _____

    b.   _____

    c.   _____

4. Write equations for these reactions.
    a.   chlorination of benzene

    b.   sulfonation of benzene, followed by treatment with NaOH

    c.   nitration of benzene

    d.   titration of phenol with NaOH

    e.   conversion of nitrobenzene to aniline

5. Draw structural formulas for these compounds.
   a. iodomethane

   b. 4-bromo-2-chlorobenzaldehyde

   c. p-nitrotoluene

   d. 6-bromo-2, 3-dichlorotoluene

ANSWERS TO SELF-TEST QUESTIONS

Multiple Choice
1. a, b, c
2. a
3. b
4. b
5. b
6. c
7. a
8. a

Completion
1. Following are structural formulas for each compound.

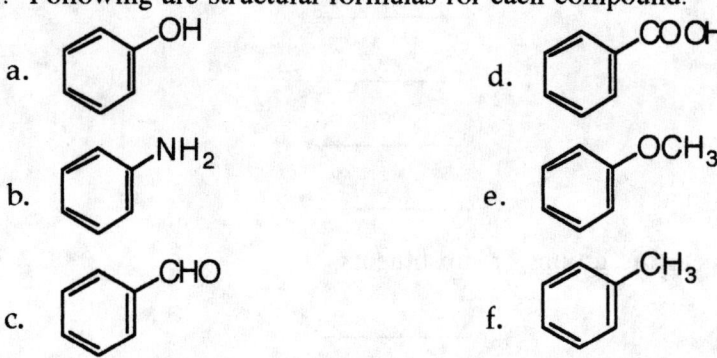

2. Following are structural formulas for the phenyl group and phenoxide ion.

3. The three compounds are
   a. o-chloronitrobenzene or 1-chloro-2-nitrobenzene
   b. m-chloronitrobenzene or 1-chloro-3-nitrobenzene
   c. p-chloronitrobenzene or 1-chloro-4-nitrobenzene

4. Following are equations for each reaction.

c. 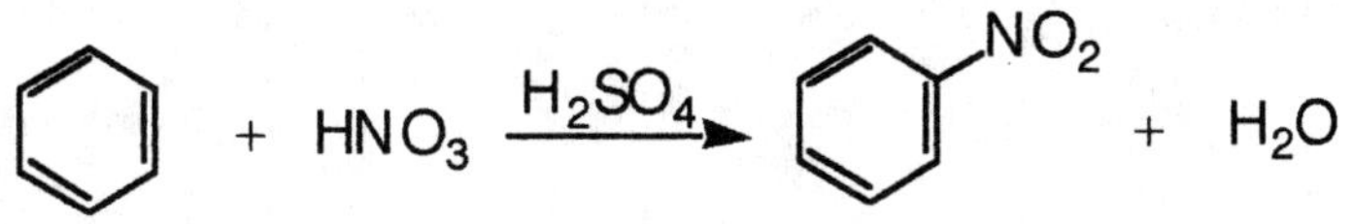

d. 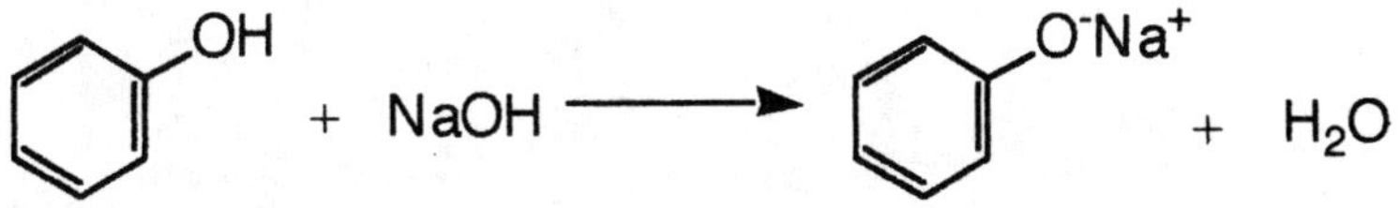

e. 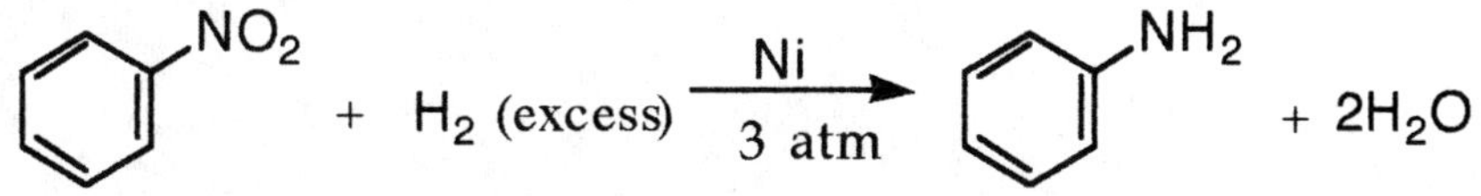

5. Following are structural formulas for each compound.

a. 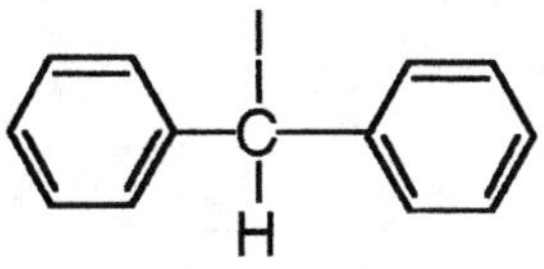

c. 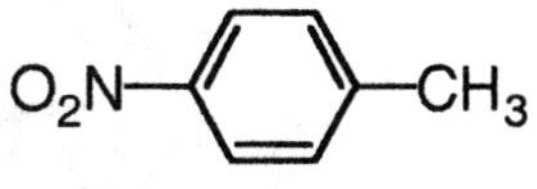

b. 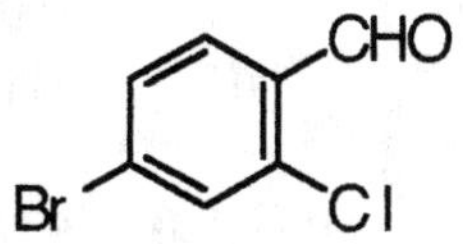

d. 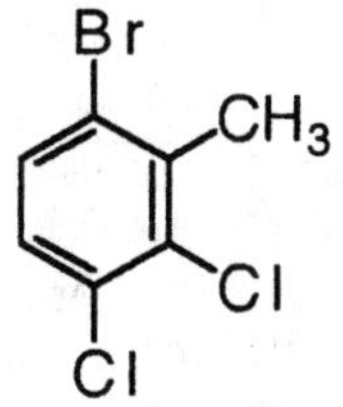

Imagination is more important than knowledge
- Albert Einstein

15

Chirality

CHAPTER OBJECTIVES

After you have studied the chapter and worked the assigned exercises in the text and the study guide, you should be able to do the following.

1. Outline the different types of isomers that have been presented in this and previous chapters and indicate an example of each.

2. Describe how connectivity (of the atoms) distinguishes stereoisomers from constitutional isomers.

3. Determine whether a molecule exhibits a plane of symmetry.

4. Determine how many stereocenters are present in a molecule and determine the maximum number of stereoisomers possible using the 2^n rule.

5. Distinguish between enantiomers and diastereomers.

6. Indicate the relationship between a racemic mixture and a pair of enantiomers.

6. Use the R, S system to assign an R or S configuration to a stereocenter.

7. Understand how plane polarized light and a polarimeter are used experimentally o characterize an enantiomer.

8. Distinguish between the terms optically active, dextrorotatory, levorotatory, and specific rotation.

9. Outline the importance of enantiomers in stereospecific interactions in living systems.

10. Define the important terms and comparisons in this chapter and give specific examples where appropriate.

IMPORTANT TERMS AND COMPARISONS

2^n Stereoisomer Rule
Chiral and Achiral Molecules
Dextrorotatory and Levorotatory
Diastereomers
Enantiomer and Enantiomeric Pair
Nonsuperposable Mirror Image
Optically Active Molecule
Plane of Symmetry
Plane Polarized Light and a Polarimeter
R, S Configuration
R, S Priorities For Groups
Racemate and Racemic Mixture
Specific Rotation

FOCUSED REVIEW

Before we finish reviewing the major functional groups in organic chemistry and embark into the subject of **biochemistry**, the important concept of chirality is presented. It will become increasingly clear that this concept, and the other chemical concepts presented in previous chapters will provide the foundation for an understanding of biochemistry. As you encounter these basic concepts, do not be shy about returning to these chapters to review and reclaim a clear understanding of any vague terms. For example, a few of the concepts and reactions that were presented in previous chapters and are relevant to this chapter are presented below. You may want to extend this summary, and prepare a similar review after each subsequent chapter.

Types of Isomers

Recall that **constitutional isomers** are molecules that have the same molecular formula, but different structural formulas (i.e., different molecular structures). As a result of the structural differences, the chemical and physical properties of the isomers are different. A simple example of this is compounds of molecular formula C_2H_6O. This formula may represent either (a) dimethyl ether or (b) ethanol.

a. CH_3OCH_3 b. CH_3CH_2OH

These two isomers have different **structures**, different **types of bonds**, different **functional groups** and correspondingly different chemical and physical properties. These isomers are often referred to as **structural isomers**.

Although structural isomers are in many cases very obvious, other isomer types are more subtle. **Stereoisomers** are compounds which have the same connections between atoms (i.e., each isomer has the same number and types of bonds), but have a different three-dimensional arrangements in space. There are two major classes of stereoisomers of interest to us.

The first type is the **cis-trans isomers** that results from the restriction of rotation about a bond. The most obvious type of isomers in this case may be the cis- and trans-isomers of a substituted olefin such as $C_2Cl_2H_2$.

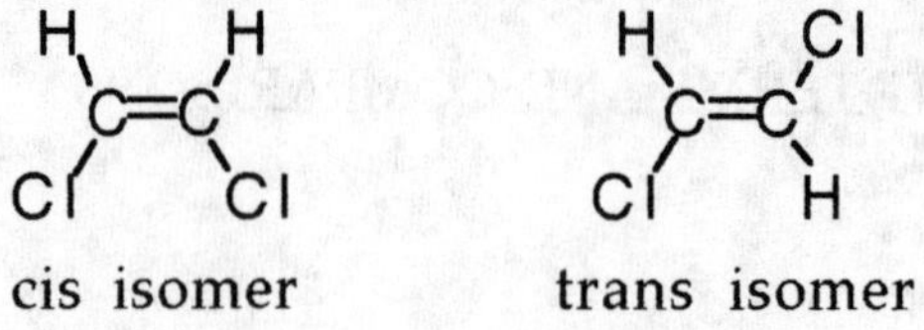

Chapter 15 Chirality

cis isomer   trans isomer

The second type is the **optical isomer,** which contain stereocenters. Plane polarized light in a polarimeter is rotated on passing through a solution of an optical isomer. Two subdivisions of optical isomers are (a) enantiomers and (b) diastereomers.

In summary, the isomeric forms of molecules can be subdivided into groups as shown below. Much of our interest in this chapter will focus on simple organic molecules that are optical isomers. However, our interest in living cells will focus interest on biological molecules, including monosaccharides, amino acids, nucleotides, lipids and the major macromolecules such as polysaccharides, proteins and nucleic acids - all of which are chiral and optically active.

Isomer Classification

Isomer Type	Character	Examples
Structural	Same molecular formula, but different molecular structure	CH_3CH_2OH CH_3OCH_3
Stereoisomers	Same connectivity of atoms, but different three-dimensional arrangement in space	
a. Cis-trans	cis- or trans-arrangement about a carbon-carbon double bond, such as observed in some fatty acids (Chapter 20)	cis- and trans-isomer of CH_2Cl_2
b. Optical		
(i) Enantiomers	Nonsuperposable mirror images	R and S forms of lactic acid
(ii) Diastereomers	Optical isomers which are not enantiomers	α- and β-forms of D-glucose

SELF-TEST QUESTIONS

MULTIPLE CHOICE. Select the correct answer from the choices listed. In some cases, more than one answer is correct.

1. Indicate which of the following molecules is chiral and how many stereocenters there are in each that is chiral.

a. 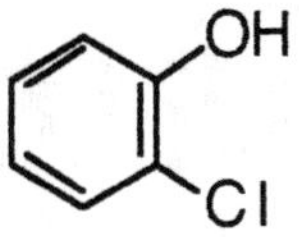 [structure: phenol with Cl ortho]

b. 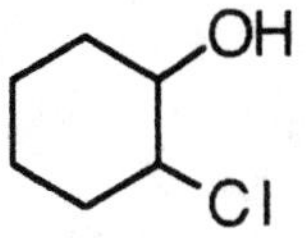 [structure: cyclohexanol with Cl]

c. 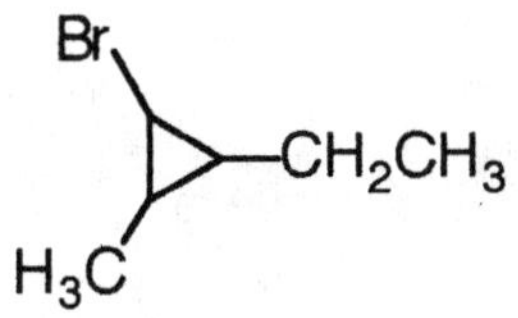 [structure: cyclopropane with Br, CH₃, and CH₂CH₃ substituents]

d. 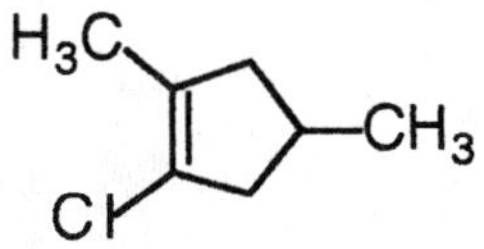 [structure: cyclopentene with H₃C, Cl, and CH₃ substituents]

e. 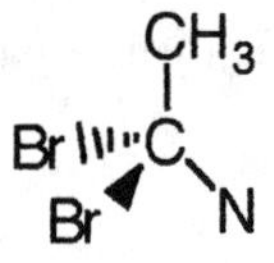 Br⁻⁻⁻C(CH₃)(Br)(N)

f. 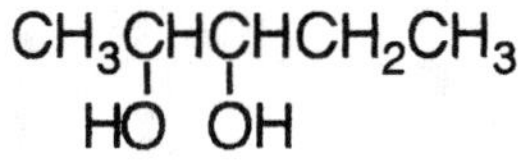 CH₃CHCHCH₂CH₃
        | |
        HO OH

2. Enantiomers have which of the following characteristics?
   a. rotate ordinary light
   b. have the same melting point
   c. have the identical molecular weight
   d. are superposable mirror images

3. A racemic mixture exhibits which of the following characteristics?
   a. does not rotate plane polarized light
   b. rotates plane polarized light
   c. contains equal concentrations of both enantiomers
   d. often contains diastereomers

4. Indicate which of the molecules in question 1 could have an enantiomer.

5. Indicate which of the following molecules are enantiomers and which ones are diastereomers.

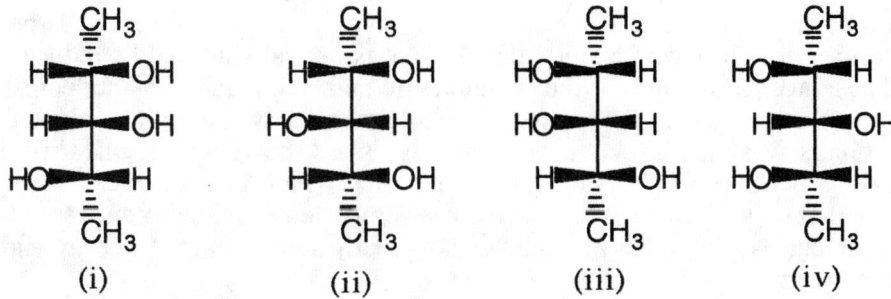

6. The priority of the groups, CH_3, COOH, NH_2 and OH in the R, S system is
   a. CH_3 > COOH > NH_2 > OH
   b. OH > NH_2 > COOH > CH_3
   c. COOH > CH_3 > NH_2 > OH
   d. COOH > OH > NH_2 > CH_3

7. Indicate whether the isomer shown is the R-form or the S-form.

Chapter 15 Chirality   106

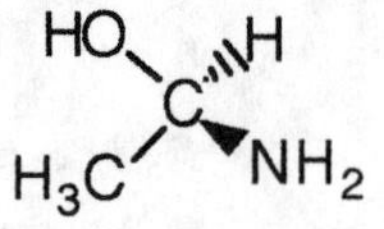

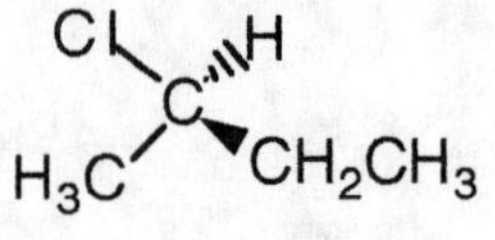

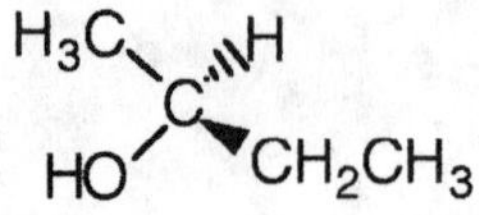

a. (structure: C with HO, H, H₃C, NH₂)
b. (structure: C with Cl, H, H₃C, CH₂CH₃)
c. (structure: C with H₃C, H, HO, CH₂CH₃)
d. (structure: C with HOCH₂, H, HOCH₂CH₂, CH₂CH₂CH₃)

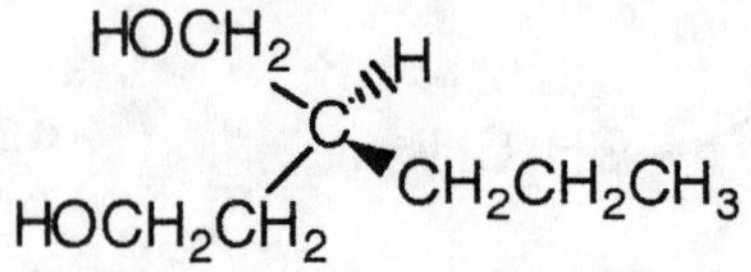

8. Calculate the maximum number of isomers in the molecules shown in questions 1 and 7.

   1a   1b
   1c   1d
   1e   1f

   7a   7b
   7c   7d

COMPLETION. Complete the following statements with the appropriate word, phrase or number.

1. Enantiomers are _____ mirror images. Because each enantiomer has at least one stereocenter, enantiomers must occur in _____.

2. The maximum number of stereoisomers that can occur for the molecule, $CH_3CH(OH)CH(OH)CH(OH)CHO$, is _____.

3. Diastereomers are _____, but they are not _____.
   While the melting points for enantiomers are the same, the melting points for diastereomers are _____.

4. Thalidomide _____ is perhaps the most notorious drug in recent history and one that is optically active. To reduce cost, the drug was sold as a racemic mixture of the enantiomeric pair. Although this antidepressant drug was not approved for use in the United States, it was widely used in Canada and Europe. One enantiomer was very effective as an antidepressant; however, pregnant women who took the drug were also subjected to the other enantiomer, which was mutagenic (causes changes in DNA) and antiabortive. As a result, many severely deformed children were born. The enantiomeric pair is shown below. Indicate how many stereocenters there are and identify the R- and the S-forms.

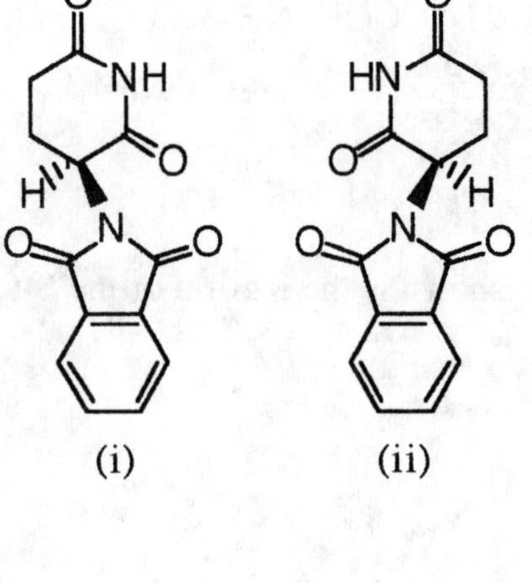

(i)   (ii)

Chapter 15 Chirality   107

ANSWERS TO SELF-TEST QUESTIONS
Multiple Choice
1.
    a.   achiral, 0   b.   chiral, 2
    c.   chiral, 3   d.   chiral, 1
    e.   achiral   f.   chiral, 2
2. b, c
3. a, c
4. all chiral molecules - b, c, d & f
5. enantiomers;   i & iii; ii & iv
   diastereomers;   i & ii; i & iv; ii & iii; iii & iv
6. b
7.   a.   R
      b.   R
      c.   S
      d.   S
8. From question 1
    a.   0   b   4
    c.   8   d.   2
    e   0   f.   4

    From question 7
    a.   2   b.   2
    c.   2   d.   2

Completion
1. nonsuperposable, pairs
2. 8
3. stereoisomers, mirror images, different
4. (i) is (S)-thalidomide; (ii) is (R)-thalidomide

"Common sense is the collection of
prejudices acquired by age eighteen"
- Albert Einstein

Amines

CHAPTER OBJECTIVES

After you have studied the chapter and worked the assigned exercises in the text and the study guide, you should be able to do the following.

1. Write a general structural formula for a primary, secondary, and tertiary amine and for a quaternary ammonium salt.

2. Write the names for simple amines, including aliphatic, cyclic, and aromatic amines.

3. Rationalize the trend in the boiling points and water solubility for simple primary amines.

4. Describe the hydrogen bonding that takes place (a) between amines and (b) between amines and water and discuss the influence of hydrogen bonding on the physical properties of amines.

5. Give the names and structures for several natural and synthetic amines that are used medicinally.

6. Write a chemical equation for the reaction of an amine with inorganic acid and with a carboxylic acid.

7. Describe how the presence of an aromatic or aliphatic group in an amine influences its base strength.

8. Define the important terms and comparisons in this chapter and give specific examples where appropriate.

IMPORTANT TERMS AND COMPARISONS

Acid/Base Chemistry of Amines
Alkaloids

Chapter 16 Amines

Amphetamines
Aromatic and Aliphatic Amine
Benzodiazepines
Heterocyclic- and Heterocyclic Aromatic Amines
Hydrogen Bonding Potential
K_b and pK_b
Organic Bases
Primary, Secondary and Tertiary Amine
Quaternary Ammonium Ion
N-Substituted Amines

SELF-TEST QUESTIONS

MULTIPLE CHOICE. Select the correct answer from the choices listed. In some cases, more than one choice will be correct.

1. The name of the compound given below is:

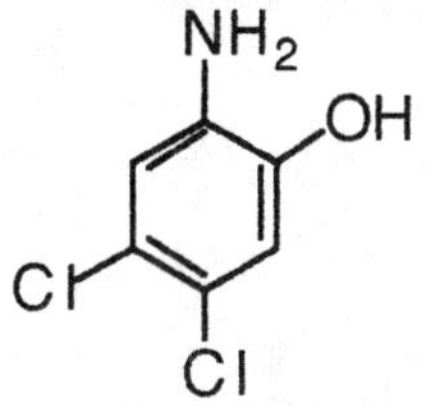

    a. 2-amino-4, 5-dichlorophenol
    b. dichlorohydroxylaniline
    c. 2-amino-4, 5-dichlorocyclohexanol
    d. 3, 4-dichloro-1-hydroxylaniline

2. The structure of N,N-methylaniline is:

    a. 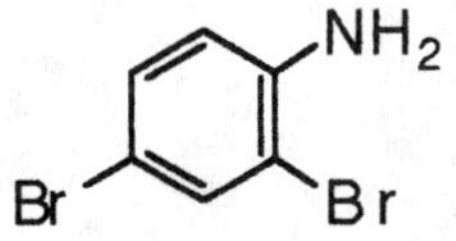
    b. 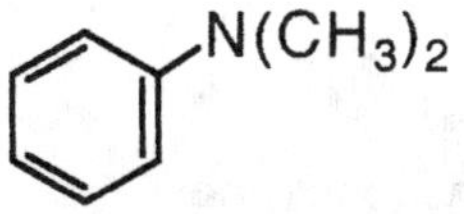
    c. 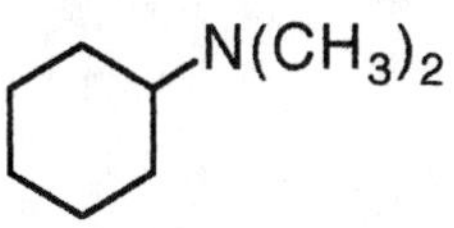
    d.

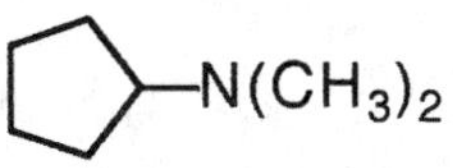

3. Protonation of amines that are used as drugs has the advantage that it
    a. increases solubility
    b. inhibits oxidation
    c. decreases reduction
    d. increases basic character

4. The pK_b or K_b value for some amines are given below. Which amine is the strongest base.
    a. $pK_b = 4.5$
    b. $K_b = 1 \times 10^{-8}$
    c. $pK_b = 7.0$
    d. $K_b = 2 \times 10^{-3}$

5. Which compound has the highest boiling point.

Chapter 16 Amines   110

    a. $CH_3CH_2CH_2OH$ c. $CH_3CH_2OCH_3$
    b. $CH_3CH_2CH_2NH_2$ d. $CH_3NHCH_2CH_3$

6. Benzodiazepine, dopamine, morphine, and procaine are all examples of:
    a.   heterocyclic ring compounds   c.   aromatic compounds
    b.   alkaloids   d.   synthetic pain killers

COMPLETION. Answer or complete the questions as indicated.
1. Indicate which compounds is an aliphatic amine, an aromatic amine, a heterocyclic amine, a aromatic heterocyclic amine, a combination of these, or none of these. Indicate whether each amine is primary, secondary, or tertiary.

    a. 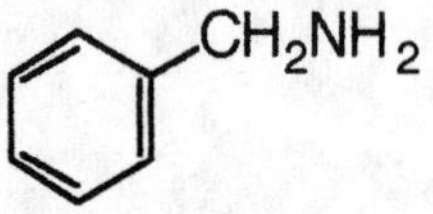 [benzyl amine structure: phenyl-CH₂NH₂]
    b. 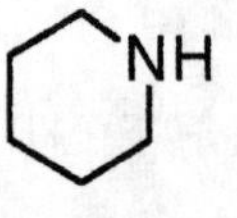 [piperidine structure with NH]
    c. 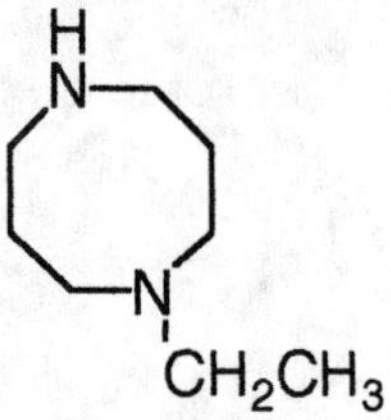 [seven-membered ring with two N atoms; one NH, other N-CH₂CH₃]
    d. 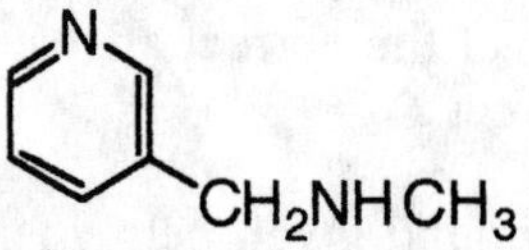 [pyridine with CH₂NHCH₃ substituent]
    e. 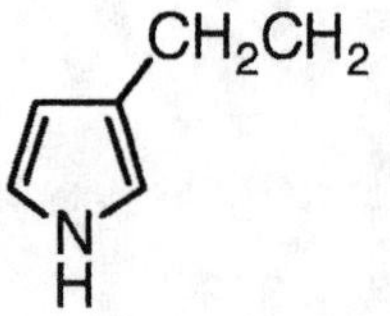 [pyrrole with CH₂CH₂ substituent]
    f. 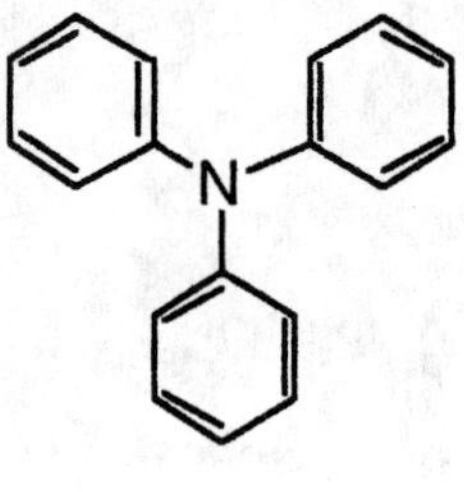 [triphenylamine structure]

2. Draw a structural formula for each compound.
    a.   ethylpropylaminej
    b.   cyclohexanamine
    c.   N-methylaniline
    d.   1-amino-4-chlorocyclohexane
    e.   5-methyl-1-heptanamine
    f.   diphenylamine

3. Name these compounds

    a. 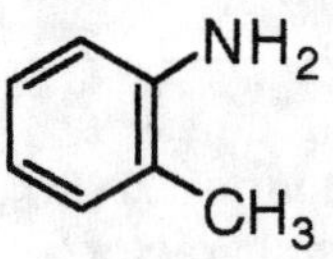 [o-toluidine: benzene with NH₂ and CH₃]
    b. 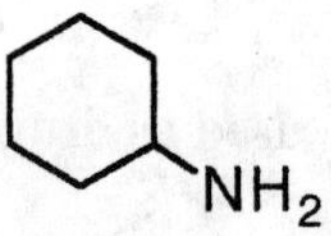 [cyclohexane with NH₂]
    c. $H_2NCH_2CH_2CH_2CH_2NH_2$
    d. 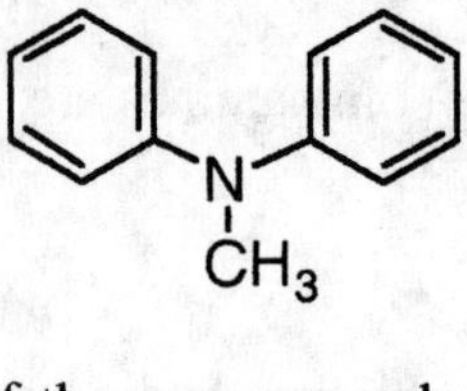 [N-methyldiphenylamine]

4. Indicate the order of increasing solubility of these compounds in water.

Chapter 16 Amines

NH$_3$, NH$_2$CH$_3$, N(CH$_3$)$_2$(C$_6$H$_5$)

_____ < _____ < _____

5. Complete the equations for the following acid/base reactions. If no reaction occurs, indicate no reaction.

   a. CH$_3$CH$_2$COOH + (CH$_3$)$_3$N $\longrightarrow$

   b. (CH$_3$CH$_2$)$_2$NH$_2$$^+Cl^-$ + NaOH $\longrightarrow$

   c. (CH$_3$)$_4$N$^+$Cl$^-$ + NaOH $\longrightarrow$

6. a. Secondary amines have _____ hydrogen(s) bound directly to the nitrogen.
   b. The compound [(CH$_3$CH$_2$)$_3$NH]$^+$Br$^-$ is a _____ and can be made by the reaction of _____ with _____.

MATCHING. Match the term in the right-hand column with the most appropriate answer in the left column.

1. Weakest class of amine bases
2. Amphetamine
3. Pyridine
4. Methylamine
5. Librium or Valium
6. Cadaverine
7. Alkaloid
8. Codeine

a. Synthetic modification of morphine
b. Tranquilizer
c. Basic nitrogen-containing compounds extracted from plants
d. Pep pills, action similar to that of epinephrine
e. End product of decaying flesh
f. Aromatic amine
g. Heterocyclic aromatic amine
h. Aliphatic amine

ANSWERS TO SELF-TEST QUESTIONS

Multiple Choice
1. a
2. b
3. a, b
4. d
5. a
6. c

Completion
1. a. a primary aliphatic amine
   b. a secondary aliphatic amine
   c. two aliphatic amines; one is a secondary and one is tertiary
   d. an aromatic amine and a secondary aliphatic amine
   e. a primary aliphatic amine and secondary aromatic amine
   f. a tertiary aromatic amine
2. Following is a structural formula for each amine.

Chapter 16   Amines

a. CH₃CH₂NHCH₂CH₂CH₃

b. 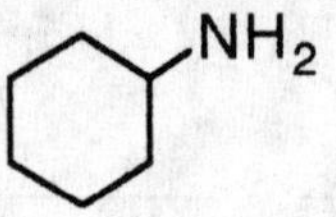 cyclohexanamine with NH₂

c. 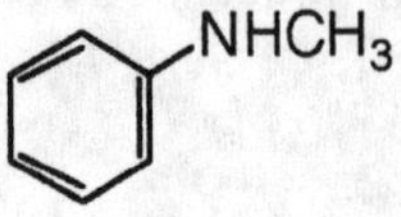 C₆H₅–NHCH₃

d. 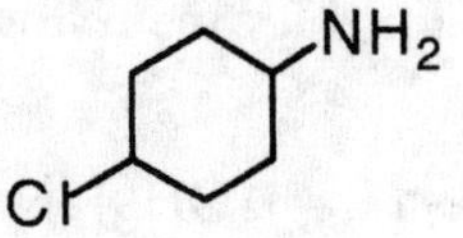 4-chlorocyclohexanamine (Cl and NH₂ on cyclohexane ring)

e. 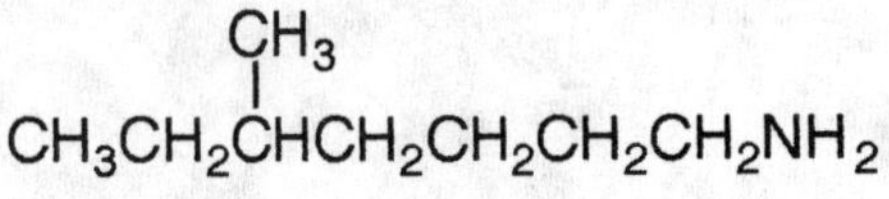 CH₃CH₂CH(CH₃)CH₂CH₂CH₂CH₂NH₂

f. 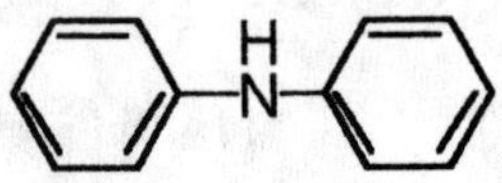 C₆H₅–NH–C₆H₅ (diphenylamine)

3. Following are the names of each amine
   a. 2-methylaniline (o-toluidine)
   b. cyclohexanamine or cyclohexylamine
   c. 1,4-butanediamine
   d. methyldiphenylamine

4. a.   N(CH₃)₂(C₆H₅)   <   NH₂CH₃   <   NH₃

5. Following are completed equations.

   a. CH₃CH₂COOH + (CH₃)₃N ⟶ CH₃CH₂COO⁻ + (CH₃)₃NH⁺

   b. (CH₃CH₂)₂NH₂⁺Cl⁻ + NaOH ⟶ (CH₃CH₂)₂NH + H₂O + NaCl

   c. (CH₃)₄N⁺Cl⁻ + NaOH ⟶ no reaction

6. a. one
   b. quaternary amine or quaternary ammonium salt, (CH₃CH₂)₃N + HBr

Matching
1. f
2. d
3. g
4. h
5. b
6. e
7. c
8. a

Chapter 17   Aldehydes and Ketones

Man's mind stretched by a new idea
never goes back to its original dimensions.
   - paraphrase from Oliver Wendell Holmes

17

Aldehydes and Ketones

CHAPTER OBJECTIVES

After you have studied the chapter and worked the assigned exercises in the text and the study guide, you should be able to do the following.

1. Write the characteristic functional group for an aldehyde and a ketone and contrast this to the structure and bond types found in functional groups presented in previous chapters.

2. Given the structural formula for an aldehyde or ketone, write an IUPAC name for the compound.

3. Given the IUPAC or common name for an aldehyde or ketone, write its structural formula.

4. Using the Table 17.1 (text) for the Order of Preference for Functional Groups, name compounds that contain two functional groups.

5. Choosing organic compounds of approximately the same molecular weight, compare the physical properties - boiling point and solubility in water - for an alkane, ketone, aldehyde, alcohol, and ether.

6. Referring back to Chapter 13, write reactions for the preparation of aldehydes and ketones that have 2, 3 and 4 carbons.

7. Write a balanced reactions for a positive Tollens' test.

8. Describe the basis for using Tollens' test to distinguish between an aldehyde and ketone.

9. Draw the general structure for a hemiacetal and an acetal which are derived from the reaction of an alcohol and an aldehyde. Provide at least one specific example of each.

10. Draw the general structure for a hemiacetal and an acetal which are derived from the reaction of an alcohol and a ketone. Provide at least one specific example of each.

Chapter 17   Aldehydes and Ketones

11. Write an equation for the formation of a hemiacetal and an acetal which are derived from a reaction of an aldehyde. Provide examples using at least two different aldehydes.

12. Write the chemical reaction for the formation of a hemiacetal and an acetal which are derived from a reaction of a ketone. Provide examples using at least two different ketones.

13. Using a single molecule which contains both an alcohol and an aldehyde or ketone, show how a cyclic hemiacetal, and then on further reaction, a cyclic acetal, is formed.

14. Write an equation for the reduction of an aldehyde or a ketone to produce an alcohol.

15. Summarize the oxidation, reduction, and addition reactions for aldehydes and ketones.

16. Define the important terms and comparisons in this chapter and give specific examples where appropriate.

IMPORTANT TERMS AND COMPARISONS
Aldehyde and Ketone
Carbonyl Group
Cyclic Acetal
Hemiacetal and Acetal (From a Ketone) Hydrolysis
Hemiacetal and Acetal (From an Aldehyde)
IUPAC Nomenclature
Keto and Enol Forms
LD_5
Order of Preference for Functional Groups
Oxidation
Primary and Secondary Alcohols
Reduction
Sufixes -al and -one
Tautomers and Tautomerism
Tollens' Test - the Silver Mirror Test

SELF-TEST QUESTIONS
MULTIPLE CHOICE. Select the correct answer or answers from the choices listed. In some cases, more than one answer is correct.

1. Mild oxidation of an aldehyde produces a carboxylic acid. However, mild oxidation of a ketone results in:
   a. an ether
   b. secondary alcohol
   c. carboxylic acid
   d. no reaction

2. Identify the compounds that are either aldehydes, hemiacetals, or acetals.

   a. [structure: diphenyl methane with OCH₃, O-phenyl, and H substituents on central carbon]

   b. $CH_3OCH_2CH_3$

Chapter 17   Aldehydes and Ketones   115

c. [structure: 4-hydroxybenzaldehyde, HO–C₆H₄–CHO]

d. [structure: methoxybenzene, C₆H₅–OCH₃]

e. CH₃CHCH₂CH
       | |
       CH₃   OCH₃
   (with OH on the CH)

f. CH₃CH₂COH
          ‖
          O

3. Select compounds in question 1 that have a carbonyl group.

4. The reaction of an alcohol with a ketone is an example of a(n):
   a. elimination reaction   c. hydrolysis reaction
   b. addition reaction   d. substitution reaction

5. Ketones are prepared by the oxidation of:
   a. secondary alcohols   c. tertiary alcohols
   b. primary alcohols   d. all of these

6. The characteristics that are produced in a positive Tollens' test are:
   a. Oxidation of an aldehyde and reduction of $Ag(NH_3)_2^+$.
   b. Reduction of an aldehyde and oxidation of $Ag(NH_3)_2^+$.
   c. Oxidation of a ketone and production of a silver mirror.
   d. All of the above.

7. The structural formula for benzaldehyde is:

   a. [C₆H₅–C(=O)–CH₃]   b. [2-hydroxybenzaldehyde]

   c. [C₆H₅–CHO]   d. [diphenyl ether]

8. In naming complex organic molecules, the order of preference for functional groups is
   a. RCOOH > RCHO > ROH > RNH₂
   b. RCHO > RCOOH > RCl
   c. RNH₂ > RCOOH > RCHO > RSH
   d. RCOOH > ROH > RCHO > RSH

9. The structural formula for 2,3-dimethylcyclohexanone is:

   a. [cyclohexanone with CH₃ at 2 and CH₃ at 3]

   b. [cyclohexanone with H₃C and CH₃ substituents]

c. 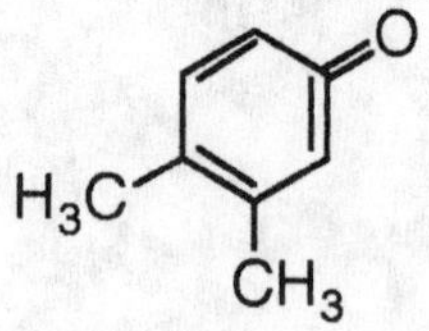 [structure: 3,4-dimethyl cyclohexadienone]   d. 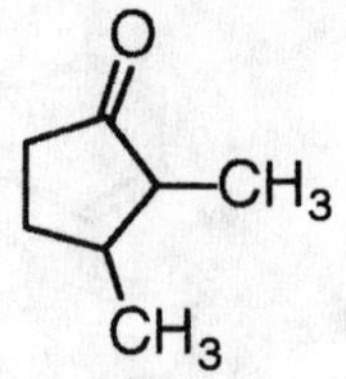 [structure: 2,3-dimethylcyclopentanone]

10. The name of this compound is:

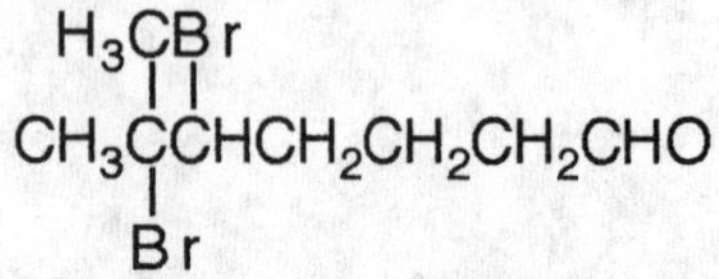

    CH₃CBr(CH₃)CHBrCH₂CH₂CH₂CHO

    a. 5,6-dibromo-6-methylheptanal
    b. 1,2-dibromo-1-methylheptanone
    c. 5,6-dibromo-1-methylheptanal
    d. 1,2-dibromo-1-methylheptanal

11. The structure for one of the forms of guanine is shown below. This form is:

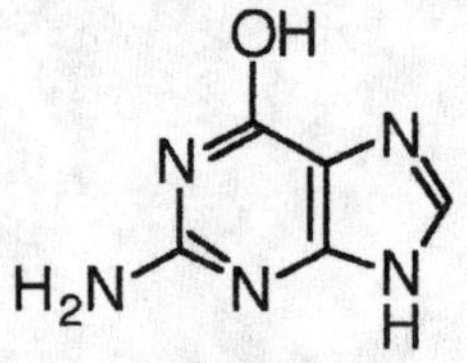

    [structure of guanine enol form]

    a. a ketone
    b. the keto form of guanine
    c. the enol form of guanine
    d. an acetal

12. In a molecule with both a carboxylic acid and an aldehyde functional group, the prefix used for the aldehyde is
    a. oxo-
    b. -ol
    c. -al
    d. -dehyde

COMPLETION. Write the word, phrase or number in the blank or draw the appropriate structure in answering the question.

1. Write the IUPAC names for the following molecules.

    a. 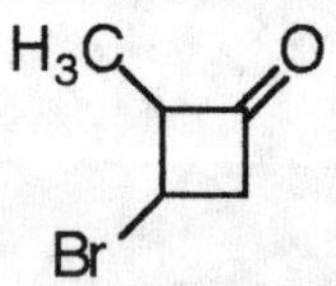 [cyclobutanone with H₃C, Br substituents]   b. 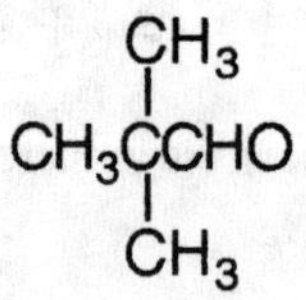 CH₃C(CH₃)(CH₃)CHO

    c. 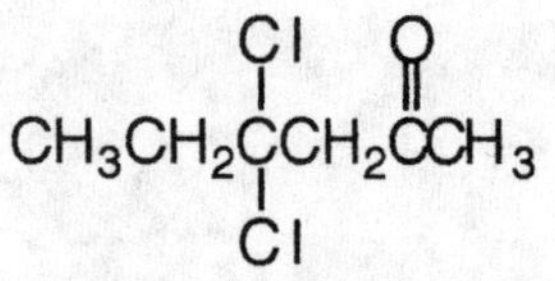 CH₃CH₂C(Cl)(Cl)CH₂COCH₃   d. 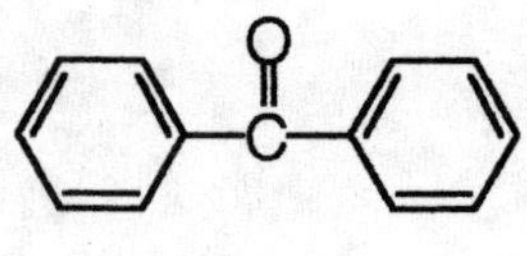 [benzophenone]

    e. 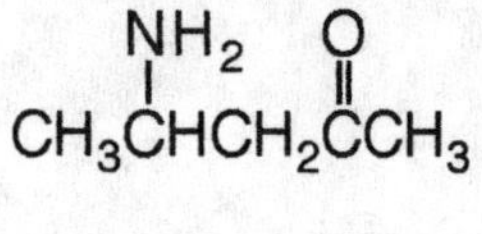 CH₃CH(NH₂)CH₂COCH₃

Chapter 17 Aldehydes and Ketones

2. Given compounds of comparable molecular weight, the order of increasing boiling points for an alcohol, alkane, ether, and aldehyde is:
   _____ > _____ > _____ > _____

3. The common name for the simplest aldehyde, CH_3CHO, is _____.

4. Complete the following equations. [H] and [O] indicate reduction and oxidation, respectively. If no reaction occurs, indicate just that.

   a. $C_6H_5\text{-}CH_2\overset{O}{\overset{\|}{C}}CH_3 \xrightarrow{[H]}$

   b. $CH_3CH_2OH \xrightarrow{[O]}$

   c. $C_6H_5\text{-}CHO + 1 CH_3OH \longrightarrow$

   d. $CH_3CH_2\underset{OCH_2CH_3}{\overset{OH}{\underset{|}{\overset{|}{C}H}}} + CH_3CH_2OH \xrightarrow{H^+}$

   e. $CH_3CH_2\overset{O}{\overset{\|}{C}}CH_3 + Ag(NH_3)_2^+ \longrightarrow$

   f. $\underset{OH}{\underset{|}{C}H_2}CH_2CH_2CH_2CHO \longrightarrow$

5. The structure of aldehydes would suggest that these compounds _____ hydrogen bond to each other.

6. The hydrolysis of an acetal yields an _____ and _____.

7. The reaction of a ketone with two molecules of an alcohol, R'OH, yields an _____ which contains two ether linkages.

8. The oxidation of $C_6H_5CH_2CHO$ yields a molecule with the structural formula
   _____

9. Compounds A and B have LD_{50} values in rat studies of 1 mg/kg and 15 mg/kg, respectively. Compound ____ is the more toxic compound.

10. The mixing of an aldehyde with an acid or base yields _____ reaction because all aldehydes are _____ acidic or basic in character.

11. Draw the keto form of compound (a) and the enol form of compound (b)

Chapter 17 Aldehydes and Ketones

compound

a. $CH_3\underset{\underset{OH}{|}}{C}=CHCH_3$

b. 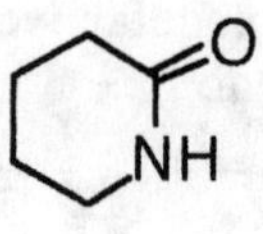 [cyclic structure: 6-membered ring with C=O and NH — δ-valerolactam]

ANSWERS TO SELF-TEST QUESTIONS

Multiple Choice
1. d
2. a (acetal), c (aldehyde), e (hemiacetal)
3. c, f
4. b
5. a
6. a
7. c
8. a
9. a
10. a
11. c
12. a

Completion
1. a. 3-bromo-2-methylcyclobutanone
   b. 2,2-dimethylpropanal
   c. 4,4-dichlorohexan-2-one
   d. diphenylketone
   e. 4-amino-2-pentanone
2. alcohol > aldehyde > ether > alkane
3. acetaldehyde
4. Following are completed equations

   a. 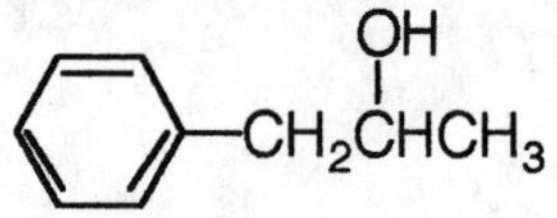 $C_6H_5-CH_2\underset{\underset{OH}{|}}{C}HCH_3$

   b. $CH_3\overset{O}{\underset{\|}{C}}H$ or $CH_3\overset{O}{\underset{\|}{C}}OH$

   c. 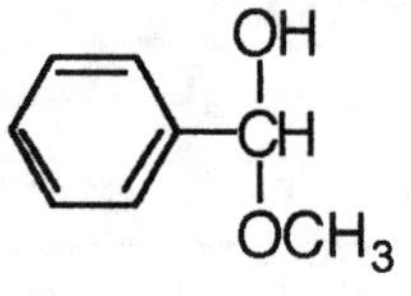 $C_6H_5-\underset{\underset{OCH_3}{|}}{\overset{\overset{OH}{|}}{C}}H$

   d. $CH_3CH_2\underset{\underset{OCH_2CH_3}{|}}{\overset{\overset{OCH_2CH_3}{|}}{C}}H$ + H_2O

   e. no reaction

   f. 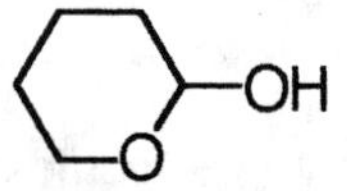 [6-membered ring with O and —OH: 2-hydroxytetrahydropyran]

5. do not
6. alcohol and aldehyde (or ketone)
7. acetal.
8. The product is a carboxylic acid, $C_6H_5CH_2COOH$
9. (compound) A
10. no, neither
11. Following is the keto form of (a) and the enol form of (b).

    a. $CH_3\overset{O}{\underset{\|}{C}}CH_2CH_3$

    b. 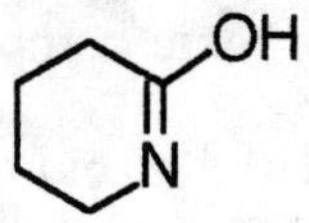 [6-membered ring with N and =C—OH group]

Chapter 18 Carboxylic Acids, Anhydrides, Esters, and Amides

18

"ASK WHY"
      - Sign in the office of Kenneth W. Kirk,
        former Associate Commissioner, FDA

Carboxylic Acids, Anhydrides, Esters and Amides

CHAPTER OBJECTIVES

After you have studied the chapter and worked the assigned exercises in the text and the study guide, you should be able to do the following.

1. Recognize the characteristic functional groups in carboxylic acids, anhydrides derived from carboxylic acids and phosphoric acid, carboxylic esters, phosphate esters, and amides.

2. Write structural formulas for these compounds, in addition to those for a carboxylate salt.

3. Draw structural formulas for the carboxylic acids containing one to six carbons; write their common names. and indicate three characteristics of carboxylic acids.

4. Explain the influence of hydrogen bonding on the physical characteristics of carboxylic acids.

5. Write a reaction for the dissociation of a carboxylic acid and also for the reaction with a strong base, such as NaOH.

6. Given the common or IUPAC name for a carboxylic acid, ester or amide, write out its structural formula.

7. Given the structural formula for a carboxylic acid, ester, or amide, name the compound by either its common or IUPAC name.

8. Contrast the physical properties of carboxylic acids with the comparable esters and amides.

9. Write an equation for the reaction of a carboxylic acid with:
    a.   a base to form a salt of a carboxylic acid
    b.   an alcohol to form an ester

Chapter 18 Carboxylic Acids, Anhydrides, Esters, and Amides   120

      c.   another carboxylic acid to form an (carboxylic) anhydride

10. Contrast the reactions and products associated with the acid- or base-catalyzed hydrolysis of an ester.

11. Write structural formulas for the anhydrides and esters of phosphoric acid and note the similarities to carboxylic anhydrides and esters.

12. Continue to write summary sheets for the reactions associated with the new functional groups presented in this chapter. These should include:
    a. Carboxylic acids and the corresponding carboxylic esters, cyclic esters, and anhydrides and the corresponding amides..
    b. Phosphoric acid and the corresponding phosphate esters and anhydrides.

13. Write reactions for the step-growth polymerization of monomers to form polyamides, polyesters, and polycarbonates.

14. Define the important terms and comparisons in this chapter and give specific examples where appropriate.

IMPORTANT TERMS AND COMPARISONS

Alkyl Diphosphates and Triphosphate
Amides and Lactams
Amines and Amides
Aspirin (Acetylsalicylic Acid)
Barbiturates
Carboxyl Group and Carbonyl Group
Carboxylic Acid Dimer
Carboxylic Acids, Esters and Anhydrides
Condensation and Step-Growth Polymer
Cyclooxygenase (COX) Enzyme
Ester and Amide Hydrolysis
Esters and Lactones
Fischer Esterification
Hydrophobic and Hydrophilic
K_a and pK_a
Kevlar, Dacron and Lexan
Le Chatelier's Principle
Penicillins and b-Lactam Ring
Phosphoric Anhydrides and Esters
Saponification
Suffixes -oic, -oate and -olactone

SELF-TEST QUESTIONS

MULTIPLE CHOICE. In the following exercises, select the correct answer from the choices listed.

1. The saponification of the ester shown below yields the following products:

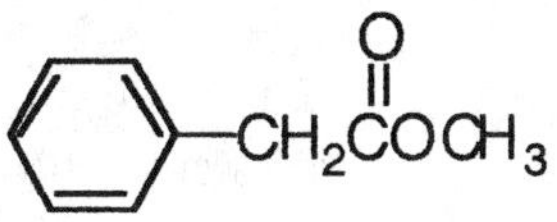

Chapter 18 Carboxylic Acids, Anhydrides, Esters, and Amides

(structure: phenyl-CH₂COOCH₃)

a. 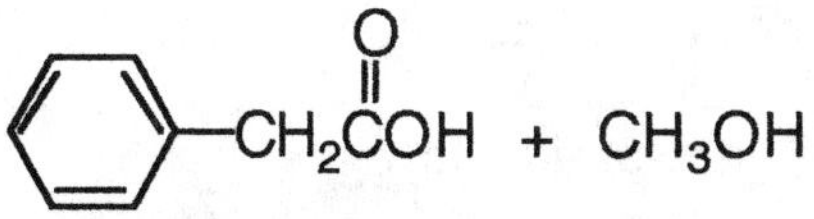 phenyl-CH₂COOH + CH₃OH
b. 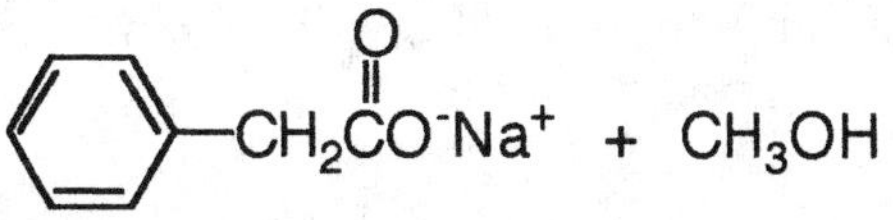 phenyl-CH₂COO⁻Na⁺ + CH₃OH
c. 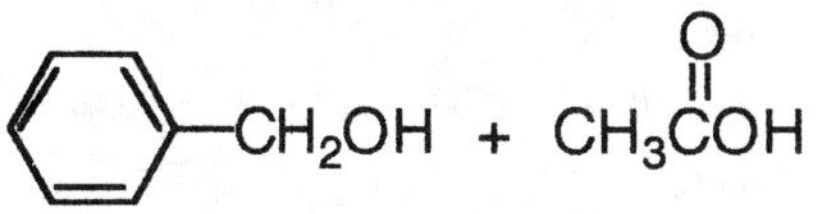 phenyl-CH₂OH + CH₃COOH
d. none of these

2. Carboxylate ion can exist in water at only:
   a. pH 7
   b. low pH
   c. high pH
   d. [H⁺] greater than 10^{-2} M

3. The name of the compound is

   $$CH_3CH_2CH_2CH_2COCH_2CH_2(CH_3)_2$$

   a. isobutyl pentanoate
   b. butyl propanoate
   c. pentyl isobutyrate
   d. isobutyl butyrate

4. Three organic acids are listed below with their pKa values.

Name	Formula	pK_a
Benzoic	C_6H_5COOH	4.19
Chloroacetic	$ClCH_2COOH$	2.86
Formic	$HCOOH$	3.74

   The order of increasing acid strength is:
   a. chloroacetic acid > formic acid > benzoic acid
   b. benzoic acid > formic acid > chloroacetic acid
   c. all are the same
   d. not determinable from available data

5. The name of this compound is

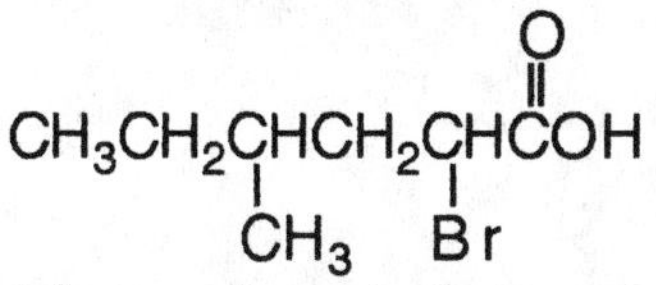

   $$CH_3CH_2CHCH_2CHCOH$$
   $$\quad\quad\quad\;\; CH_3 \;\; Br$$

   a. 1-bromo-3-methylhexanoic acid
   b. 2-bromo-4-methylhexanoic acid
   c. 2-bromo-4-methylpentanoic acid
   d. 5-bromo-3-methylpentanoic acid

6. The hydrolysis of what ester produces butanoic acid and phenol.
   a. phenyl butanoate
   b. phenyl propanoate
   c. butyl benzoate
   d. phenyl propanoate

Chapter 18 Carboxylic Acids, Anhydrides, Esters, and Amides

7. 5-Pentanolactam is an example of
   a. an ester
   b. an amide
   c. a cyclic ester
   d. a cyclic amide

8. The ethyl group in N-ethylbutanamide is bonded directly on the
   a. carbonyl carbon
   b. carbon 3
   c. carbon 2
   d. the amide nitrogen

COMPLETION. Write the word, phrase or number in the blank or draw the appropriate structure in answering the questions.

1. Write out the functional group or groups contained in each of the following condensed structural formulas. Some of the functional groups were presented in previous chapters.

   a. $CH_3CH_2CH=CHCOOH$ _____

   b. $CH_3\overset{O}{\overset{\|}{C}}CHCH_3\overset{OH}{\underset{CH_3}{\overset{|}{C}}}CH_3$ _____

   c. (benzene ring with SH and CHO substituents) _____

   d. (benzene ring with $-\overset{O}{\overset{\|}{C}}O\overset{O}{\overset{\|}{C}}CH_3$) _____

   e. $CH_3CH_2O\overset{O}{\overset{\|}{\underset{OH}{P}}}OH$ _____

   f. $(CH_3)_2CHO\overset{O}{\overset{\|}{C}}CH_3$ _____

   g. (benzene ring with COOH and $O\overset{}{\underset{O}{\overset{\|}{C}}}CH_3$) _____

   h. $CH_3COO^-Na^+$ _____

2. Complete these equations and name the type of reaction.

   a. _____ + NaOH $\longrightarrow$ $CH_3COO^-Na^+$ + H_2O

   b. (phenyl)—COOH + CH_3CH_2OH $\xrightarrow{H^+}$ _____

   c. _____ $\xrightarrow{H_2CrO_4}$ $(CH_3)_2CHCH_2COOH$

   d. CH_3CH_2COOH + _____ $\longrightarrow$ $CH_3\overset{O}{\overset{\|}{C}}O\overset{O}{\overset{\|}{C}}CH_2CH_2CH_3$

Chapter 18 Carboxylic Acids, Anhydrides, Esters, and Amides

e. C$_6$H$_5$-COOH ⇌ _____ + H$^+$

f. _____ + _____ $\xrightarrow{H_2O}$ CH$_3$CH$_2$COOH + NH$_4$Cl

3. The structure of the dimer formed by hydrogen bonding between 2 carboxylic acid molecules, RCOOH, is:
   _____

4. Polyesters are examples of condensation polymers. To prepare these large molecules, each reactant molecule must have at least two functional groups. Draw out at least one repeating unit produced in the polyester formed in the reaction shown.

   HOCH$_2$CH$_2$OH + HOOCCH$_2$CH$_2$CH$_2$COOH ⟶ _____

5. The order of increasing boiling points for an alkane, a carboxylic acid, an ester, and an alcohol, all of comparable molecular weight, is:

   _____ > _____ > _____ > _____

6. The primary reason for the order observed in the compounds in question 5 is _____
   _____.

7. The structural formula for octyl acetate is CH$_3$COO(CH$_2$)$_7$CH$_3$. This molecule gives the fragrance of an orange. Draw the structure of the alcohol and the acid from which the ester was made.

   _____   _____
   alcohol   acid

8. Name the following compounds.

   a. 3-Br-C$_6$H$_4$-COOH _____

   b. CH$_3$CO-O-C$_6$H$_5$ _____

   c. CH$_3$CH$_2$CCl$_2$COO$^-$ K$^+$ _____

Chapter 18 Carboxylic Acids, Anhydrides, Esters, and Amides 124

9. Draw the structures for the following molecules.
   a. 2-phenyl propanoic acid
   b. phenyl 2-methyl propanoate
   c. isopropyl benzoate
   d. 6-hexanolactone
   e. N-methylpropanamide
   f. ammonium pentanoate
   g. methyl 3-phenylpentanonate

10. Aspirin is now known to _____ the synthesis of _____ by reacting with the enzyme, _____, with the acronym, COX.

ANSWERS TO SELF-TEST QUESTIONS
Multiple Choice
1. c
2. c
3. a
4. a
5. b
6. a
7. d
8. d

Completion
1. a. alkene, carboxylic acid
   b. ketone, alcohol (note that this is not a carboxylic acid)
   c. thiol, aldehyde
   d. anhydride
   e. phosphate (mono) ester
   f. ester
   g. carboxylic acid, ester
   h. sodium salt of a carboxylic acid

2. a. CH_3COOH, neutralization of a carboxylic acid

   b. 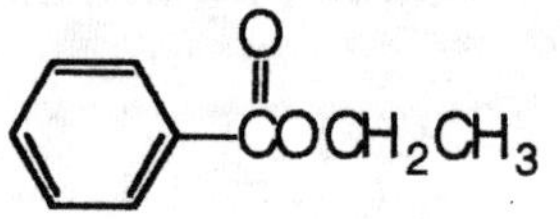 Ph—COOCH$_2$CH$_3$ + H$_2$O   Fischer esterification; preparation of an ester

   c. $(CH_3)_2CHCH_2CH_2OH$ or $(CH_3)_2CHCH_2CHO$ oxidation of a primary alcohol or an aldehyde

   d. CH_3COOH preparation of an anhydride

   e. 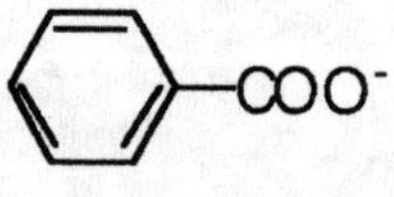 Ph—COO$^-$

   f. $CH_3CH_2CONH_2$ + HCl

3. The dimer has two hydrogen bonds.

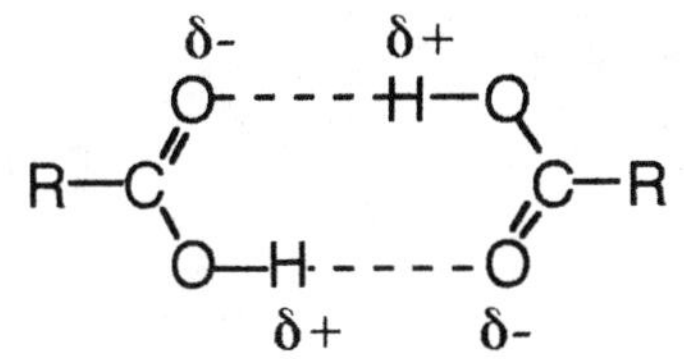

Chapter 18 Carboxylic Acids, Anhydrides, Esters, and Amides   125

4. The following structure shows two repeating units of the polymer. The units derived from ethylene glycol are circled.

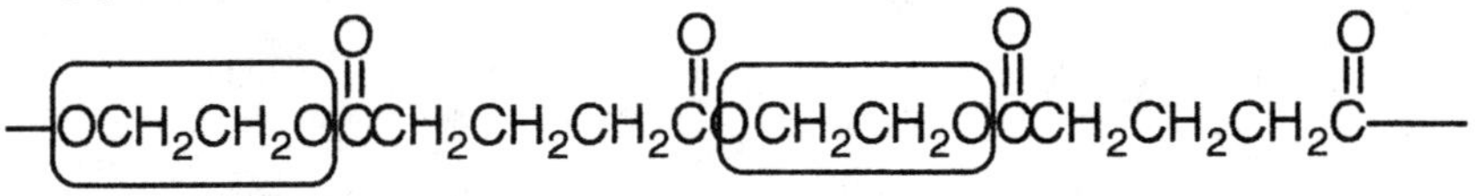

5. carboxylic acid > alcohol > ester > alkane

6. hydrogen bonding between molecules. The greater the hydrogen bonding between molecules, the more energy it requires to vaporize the liquid and therefore the higher is the boiling point.

7. $CH_3(CH_2)_7OH + CH_3COOH$

8. a. 3-bromobenzoic acid
    b. phenyl acetate
    c. potassium 2,2-dichlorobutanoate

9. Following is a structural formula for each compound.

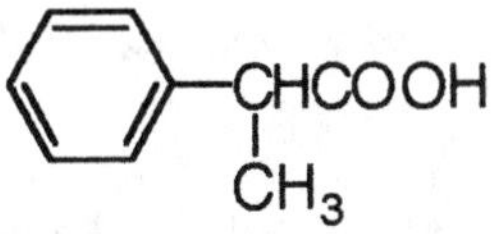

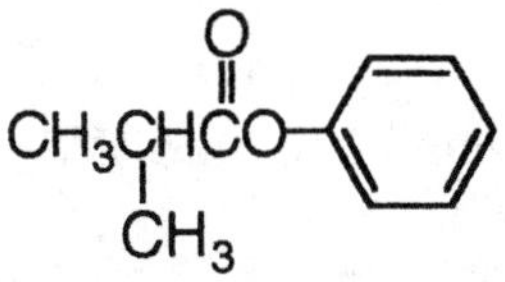

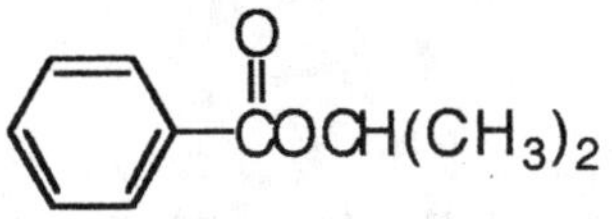

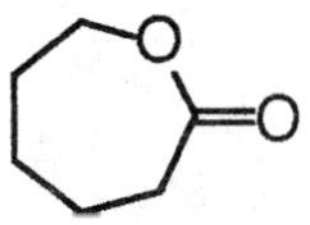

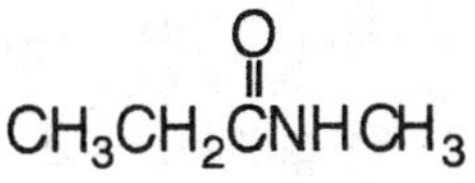

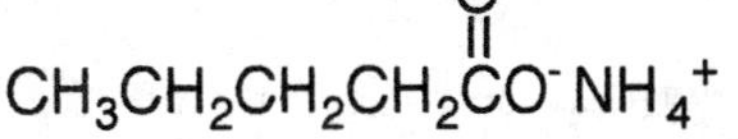

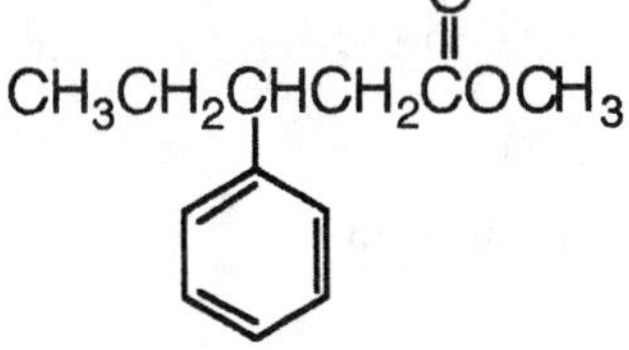

10. inhibit, prostaglandins, cyclooxygenase

Burnt Sugar Icing Recipe
 2/3 cup granulated sugar
 1/3 cup boiling water
 3 cups powdered sugar
 2 tablespoons butter
   - H. Morgan
      You Can't Eat That!

Carbohydrates

CHAPTER OBJECTIVES

After you have studied the chapter and worked the assigned exercises in the text and the study guide, you should be able to do the following.

1. Define carbohydrate and indicate a number of specific uses of carbohydrates in bacteria, plants and humans.

2. Draw the D-form of the aldoses and ketoses that have three, four, and five carbon atoms. Identify all stereocenters and the position which defines the molecule as being either the D- or the L-form.

3. Calculate the maximum number of stereoisomers possible for aldoses and ketoses with 3, 4 and 5 carbon atoms. Draw the Fischer projection formulas for these and identify the enantiomeric pairs of molecules, and the diastereomers. Specify the D- and L-form of these monosaccharides.

4. Indicate the physical and chemical properties that differ in enantiomers.

5. List four properties that are identical for enantiomers.

6. Indicate the relationship between a racemic mixture and a pair of enantiomers.

7. Describe in words the solution equilibrium for a monosaccharide such as β-D-glucose. Also draw the open chair and ring structures for the molecules in the equilibrium. Identify the hemiacetal units in the Haworth formulas.

Chapter 19 Carbohydrates

8. Write the reactions for D-glucose undergoing (a) reduction, (b) oxidation, and (c) phosphorylation.

9. List three monosaccharides and three disaccharides of biological importance. Name the products obtained from the hydrolysis of the each disaccharide.

10. Draw the Haworth structure for maltose, pointing out (a) the acetal unit, (b) the reducing ring and (c) the glycosidic linkage.

11. Describe a difference in structure between amylopectin and amylose. Draw a minimum unit of this structure to show the difference.

12. Describe the major structural difference between starch and cellulose. Explain the significance of this with regard to human metabolism.

13. Characterize glycosidic bonds found in the hyaluronic acid and heparin. Explain why both these polysaccharides are considered acidic.

14. Define the important terms and comparisons in this chapter and give specific examples where appropriate.

IMPORTANT TERMS AND COMPARISONS

2^n Isomers
1,4- and 1,6-Glycosidic Bonds
A, B, AB and O Blood Groups

α- and β-D-Glucose
Acidic Polysaccharide
Alditols and Sorbitol
Aldose and Ketose
Amylose, Amylopectin and Glycogen
Anomer and Anomeric Carbon Atom
Antigen and Antibody
Ascorbic Acid (Vitamin C)
Cellulose and Cellulase
Chair Conformation
Chair Conformation
Chiral Molecule
Chiral Molecule
D- and L-Glucose
Diabetes Mellitus
Enantiomer and Diastereomer
Enediol
Fischer Projection and Haworth Formula
Glucose Oxidase
Glucose, Dextrose, or Blood Sugar
Glycoside and Glucoside
Hemiacetal and Acetal
Hyaluronic Acid and Heparin
Hydrolysis and Condensation Reactions
Hypo- and Hyperglycemia
"Locked" Ring of a Disaccharide

Chapter 19 Carbohydrates

Mono-, Di-, Oligo- and Polysaccharides
Mutarotation
Nonsuperposable Mirror Image
Osmotic Pressure
Photosynthesis
Pyranose and Furanose Rings
Racemate and Racemic Mixture
Reducing and Non-Reducing Sugar
Saccharin
Starch and Cellulose
Sucrose, Lactose and Maltose
Uronic Acids

SELF-TEST QUESTIONS

MULTIPLE CHOICE. Select the correct answer from the choices listed. In some cases, more than one answer is correct.

1. Which of the following properties are associated with D-fructose?
   a. optically active
   b. undergoes mutarotation
   c. disaccharide
   d. rotates plane-polarized light
   e. is or contains a reducing sugar
   f. contains an α-1,4-glycosidic bond

2. Enantiomers have which of the following characteristics?
   a. rotate ordinary light
   b. have the same melting point
   c. react with optically active molecules at the same rate
   d. are superposable mirror images

3. Starch has which of the following characteristics?
   a. β–1,4 glycosidic bonds
   b. made up of amylose and glycogen
   c. is a monosaccharide
   d. optically active

4. Which of the following molecules are disaccharides?
   a. heparin
   b. sucrose
   c. maltose
   d. acetone

5. Although the Haworth formula is a simple and useful representation of a cyclic sugar, in many ways, it does not correctly represent the bond angles about the carbon atoms in the sugar molecule. These bond angles are:
   a. 120°
   b. 109.5°
   c. 150°
   d. 90°

6. The oxidation of D-glucose by glucose oxidase to produce D-gluconic acid occurs at which carbon atom?
   a. C-1
   b. C-2
   c. C-6
   d. C-4

Chapter 19   Carbohydrates   129

7. The IUPAC name for D-ribose, an aldopentose, is
   a. 1, 2, 3, 4-tetrahydroxypentanol
   b. 2, 3, 4, 5- tetrahydroxypentanal
   c. 1, 2, 3, 4, 5-pentahydroxypentanal
   d. 1-tetrahydroxypentanal

8. D-glucose can be converted to ascorbic acid by some plants and animals, but not humans. Ascorbic acid contains a number of functional groups, but does not contain
   a. lactone
   c. aldehyde
   b. chiral carbon
   d. alcohol functional group

COMPLETION. Complete the following statements with the appropriate word, phrase or number.

1. The process in which the α- and β-ring forms of a monosaccharide interconvert by way of the open chain structure is called _____.

2. Photosynthesis is the process in which plants combine _____ and _____ to produce _____. Both _____ and _____ are required for this reaction.

3. The hydrolysis of sucrose produces _____ and _____.

4. The polysaccharide which contains the β-1,4 glycosidic linkage is _____.

5. The Haworth projection of glucose does not indicate that the glucose is actually in a _____ conformation, with the hydrogens or hydroxyl groups directed in either _____ or _____ positions.

6. A pair of diastereomers is _____ which are not _____. The melting points for these two isomers are _____.

7. Amylopectin differs from amylose in that it contains _____ which involves _____ glycosidic bonds.

8. Blood type _____ is the universal donor because it has _____ of the blood _____ A or B.

9. The number of stereocenters in an aldoheptose is _____; therefore, the maximum number of stereoisomers possible is _____.

10. Lactose is an example of a _____ disaccharide because it contains a _____ unit; therefore it (will or will not) undergo mutarotation.

11. Two common features of heparin and hyaluronic acid are that they both contain _____ and _____.

12. The _____ form of tartaric acid is optically _____ and is a _____ of the enantiomeric pair ((+) and (-) forms) of isomers.

Chapter 19 Carbohydrates

13. The number one carbon in glucose is the called the _____ carbon. It is this position which determines whether the D-glucose is in the _____- or the _____- form. These two forms of glucose are not enantiomers, but are _____.

14. _____ is the name of the disaccharide that contains _____ glucose residues, in which a carbon on each glucose is linked together by an _____ (a functional group) linkage. This glycosidic bond is between carbon number _____ on one glucose and carbon number _____ on the other glucose. In the reaction of any the two sugar molecules to yield the disaccharide, one molecule of _____ is also produced.

15. Humans cannot digest cellulose because they do not have the enzyme, _____, which cleaves a _____ bond in cellulose.

ANSWERS TO SELF-TEST QUESTIONS

Multiple Choice
1. a, b, d, e
2. b
3. d
4. b, c
5. b
6. a
7. b
8. c

Completion
1. mutarotation
2. H_2O, CO_2, carbohydrates; light ; chlorophyll
3. glucose, fructose
4. cellulose
5. chair, axial, equatorial
6. optical isomers, enantiomers, different
7. chain branching, (1-6)
8. 0, neither, antigens
9. 5, $2^5 = 32$
10. reducing, hemiacetal, will
11. pyranose rings, acidic side groups
12. meso, inactive, diastereomer
13. anomeric, α–, β–, diastereomer
14. Maltose, 2, ether, 1, 4, water
15. cellulase, β-1,4-

It takes a membrane to make sense
out of disorder in biology.
- Lewis Thomas
   The Lives of a Cell

20

Lipids

CHAPTER OBJECTIVES

Lipids are defined **operationally** - that is, those molecules in a cell that can be extracted with non-polar solvents are collectively classified as lipids. This classification procedure is not very rigorous and as a result a rather diverse group of molecules, with very different functions are enveloped within the definition. The focus of this chapter is to define the (1) COMPOSITION, (2) STRUCTURE, (3) PROPERTIES and (4) FUNCTION of these molecules in the cell. Keep this in mind as you review the material. As you will see, the functions range from providing the basic structure of all membranes, to highly specific hormones, some of which regulate metabolism, while others are very specific and define our sex.

After you have studied the chapter and worked the assigned exercises in the text and the study guide, you should be able to do the following.

1. Name the four main subclasses of lipids and give a few examples of each.

2. Draw the general structure of a mono-, di- and triglyceride (all fats) which includes saturated and/or unsaturated fatty acids.

3. Explain how the unsaturation in fatty acids influences their melting point.

4. Write out the general equations for the (1) hydrogenation of unsaturated fats and oils and for the (2) saponification of all fats or oils. Outline the relationship between cholesterol, progesterone and the adrenocorticoid hormones.

5. Define the lipid bilayer and describe the essential features of the fluid mosaic model of a membrane.

6. Draw a phosphoglyceride and a sphingolipid and indicate their main function.

7. Draw the structure for cholesterol and outline the relationship between cholesterol, progesterone and the adrenocorticoid hormones.

8. Outline the interrelationship between HDLs, LDLs, cholesterol and its metabolism, and the important roles that each plays in the development of arteriosclerosis.

Chapter 20 Lipids

9. Outline the details of how bile salts and lipases work in concert to break down lipids.

10. Draw the structure of arachidonic acid and point out some common structural changes that occur on production of prostaglandins, leukotrienes and thromboxanes.

11. Define the important terms and comparisons in this chapter and give specific examples where appropriate.

IMPORTANT TERMS AND COMPARISONS

Hydrophobic and Hydrophilic
Fat, Oils and Waxes
Glycerol
Fatty Acid
Van der Waals Force
Ester Linkage
Cis Isomer
Essential Fatty Acid
Polyunsaturated
Hydrogenation
Rancidity
Saponification
Soap and Detergent
Membrane and Lipid Bilayer
Phospholipid and Glycolipid
Phosphoglyceride
Choline
HDLs and LDLs
Micelles
RU486 and Progesterone
COX-1 and COX-2 Enzymes
Myelin Sheath

Phosphatidylcholine or Lecithin
Cephalin and Lecithin
Sphingosine
Sphingomyelin
Tay-Sachs Disease
Steroid and Cholesterol
Progesterone
Mineralcorticoid
Glucocorticoid
Aldosterone and Cortisol
Bile Salts
Sex Hormones
Testosterone
Estradiol
Oral Contraceptives
Lipase
Prostaglandins, Leukotrienes and Thromboxanes
Lovastatin and HMG-CoA Reductase
Arteriosclerosis and Myocardial Infarction
Active, Facilitated and Passive Transport
Rancidity
Steroid and Nonsteroid Anti-Inflammatory Drugs (NSAIDS)

SELF-TEST QUESTIONS

MULTIPLE CHOICE. In the following exercises, select the correct answer from the choices listed. In some cases, two or more choices will be correct.

1. Fats differ from waxes in that fats have:
   a. more unsaturation
   b. a glycerol backbone
   c. higher melting points
   d. longer fatty acids

2. Oleic acid is an unsaturated fatty acid containing:
   a. 12 carbons
   b. 14 carbons
   c. 16 carbons
   d. 18 carbons

3. Both stearic acid and linoleic acid have 18 carbons; however, linoleic acid is unsaturated, while stearic acid is saturated. The melting point of stearic acid:
   a. is higher
   b. is lower
   c. is the same as linoleic acid
   d. relative to that for linoleic acid cannot be predicted

Chapter 20 Lipids

4. Liquid soaps are generally salts of fatty acids in which the cation is:
   a. K^+
   b. Na^+
   c. Ca^{2+}
   d. Mg^{2+}

5. Which of the following structures represents cholesterol?

   a.

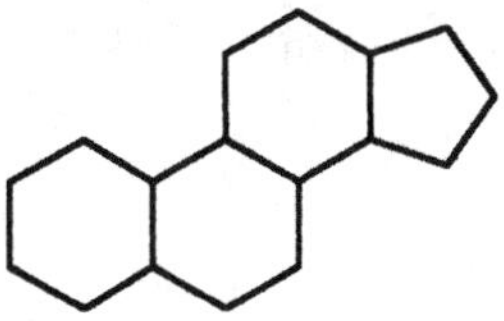

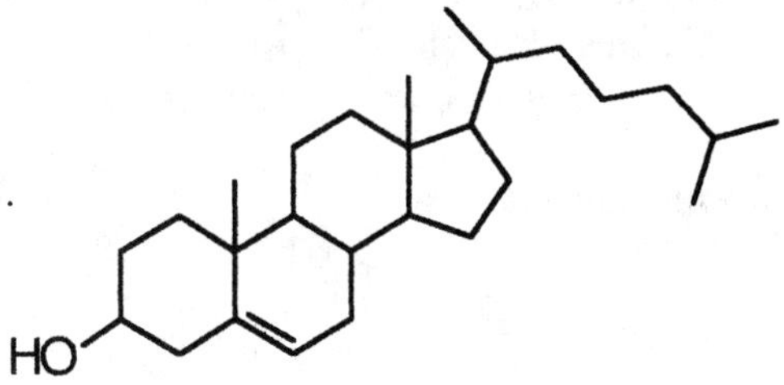

   b.

   c.

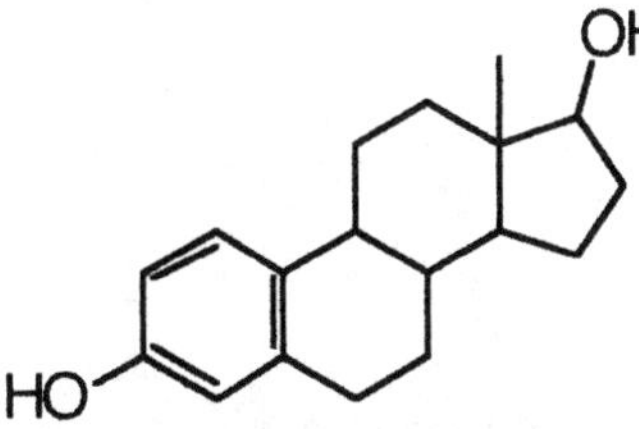

   d.

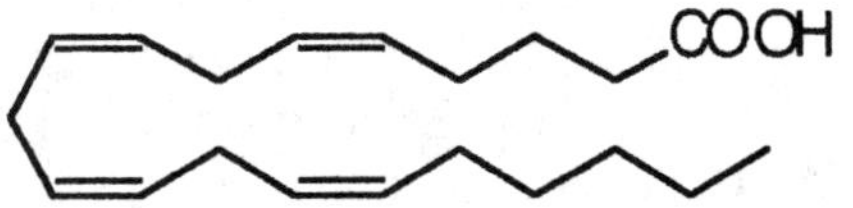

6. Prostaglandin group, PGE$_2$, possesses:
   a. ring closure at C-8 and C-12
   b. 20 carbon atoms
   c. a carbonyl group at C-9
   d. two double bonds

7. Multiple sclerosis is thought to be associated with:
   a. degradation of the myelin sheath
   b. inhibited prostaglandin synthesis
   c. lipid storage
   d. a deficiency of enzyme COX-1

8. Lipid bilayers are composed mainly of:
   a. complex lipids
   b. steroids
   c. prostaglandins
   d. fatty acids

9. Basic hydrolysis or saponification of triglycerides will produce:
   a. glycerol
   b. fatty acids
   c. salts of fatty acids
   d. bile salts

10. The 18-carbon backbone of sphingosine is found in:
    a. sphingomyelin
    b. myelin sheath
    c. phosphatidylcholine
    d. testosterone

11. Oils are:
    a. liquid fats
    b. completely saturated
    c. high molecular weight fatty acids
    d. colored

Chapter 20 Lipids

12. Derivatives of what molecules are effective oral contraceptives?
    a. progesterone
    b. estradiol
    c. PGF_{2a}
    d. cortisone

13. Low density lipoprotein (LDL) has which of the following characteristics:
    a. Contains more cholesterol than high density lipoprotein (HDL).
    b. Soluble in the blood plasma
    c. Carries cholesterol to the cells
    d. Binds to receptor molecules on the cell surface within the coated pits.

14. Anabolic steroids are
    a. powerful female sex hormones
    b. very effective detergents that aid digestion
    c. effective in the treatment of head aches
    d. used by athletes to increase muscle development

15. The fluid mosaic model of the membrane is characterized by
    a. a lipid bilayer made up of complex lipids
    b. proteins embedded in the bilayer
    c. cholesterol molecules inserted in the bilayer
    d. all of these characteristics

16. Derivatives of this group of membrane lipids function as signaling molecules in chemical communication.
    a. cephalins
    b. ceramides
    c. phosphatidyl inositol
    d. glucocerebrosides

COMPLETION. Write the word, phrase or number in the blank space in answering the question.

1. Energy is stored in our bodies in the form of _____ because burning it produces twice as much energy as the burning of _____.

2. Aspirin has been found to inhibit _____ synthesis.

3. Both the physiologically potent prostaglandins and leukotrienes are derived from the 20-carbon, unsaturated fatty acid called _____ _____.

4. The enzymes involved in lipid metabolism are called _____.

5. A widely known disease associated with lipid storage is _____.

6. _____ are negatively charged salts, which are powerful detergents used in the digestion of lipids. They are derived from the oxidation of _____.

7. The hydrophobic segment of a detergent is associated with the _____.

8. All unsaturated fatty acids exist as _____ at room temperature.

9. Rancidity is caused in part by the _____ of the fat or oil to aldehydes or by the production of _____ by the hydrolysis of triglycerides.

10. The fluid mosaic model of membranes proposed that _____ molecules are floating on and within the _____.

Chapter 20 Lipids   135

11. Fats are chemically the equivalent of an _____ (a functional group), composed of _____ and ____ _____ with an even number of carbons. Waxes are similar, but are usually always _____ (liquid or solid) because the alcohol component is a _____ _____ alcohol.

12. The common name for phosphatidylcholine is _____.

13. Leukotrienes are _____ of hormonal responses. Although similar in structure to prostaglandins, there is no _____ _____ in the structure of leukotrienes.

14. A faulty cholesterol transport system is often characterized by high levels of _____ and low levels of _____, many times resulting from not enough functional. _____ _____ (two words)..

15. Aldosterone is a _____, which regulates the concentrations of _____ and is synthesized in the _____ gland.

16. The only steroid hormone that contains an aromatic ring is _____.

17. A student once told me that the way he remembers the steroid skeleton is that he compares it to a house that contains "three rooms and a bath". In this analogy, the three rooms refer to the three _____ and the bath to the _____.

ANSWERS TO SELF-TEST QUESTIONS

Multiple Choice
1. b
2. d
3. a
4. a
5. c
6. a, b, c, d
7. a
8. a
9. a, c
10. a
11. a
12. a
13. a, b, c, d
14. d
15. d
16. c

Completion
1. fat, carbohydrates
2. prostaglandin
3. arachidonic acid
4. lipases
5. any of the 5 listed in Box 20F in the text
6. Bile salts, cholesterol
7. non-polar hydrocarbon chain
8. liquids
9. oxidation, fatty acids
10. protein, lipid bilayer
11. esters, glycerol fatty acids, solid, long chain
12. lecithin
13. mediators, ring closure
14. LDL, HDL, LDL receptors
15. mineralocorticoid, ions, adrenal
16. estradiol
17. 6-membered rings, 5-membered ring

21

"Isn't it astonishing that all these secrets
have been preserved for so many years
just so that we could discover them."
- Orville Wright
Miracle at Kitty Hawk

Proteins

CHAPTER OBJECTIVES

This chapter, similar to the previous chapter on lipids, concentrates on the (1) COMPOSITION, (2) SEQUENCE, (3) the levels of STRUCTURE and the (4) diverse FUNCTIONS of proteins. Proteins and/or enzymes are associated with, in some way, virtually all reactions in the cell and therefore will be of interest in all aspects of cellular regulation and metabolic processes.

After you have studied the chapter and worked the assigned exercises in the text and the study guide, you should be able to do the following.

1. List eight major functions of proteins and give an example of a protein that is involved in each biological function.

2. Write out the structures for at least two amino acids which are representative of an amino acid with a (1) nonpolar, (2) polar, but neutral, (3) acidic and (4) basic R group. Also indicate the unique character of the amino acids glycine, proline and cysteine.

3. Demonstrate the amphoteric character of amino acids by writing the reactions for the zwitterion with an (1) acid or (2) base.

4. Write out the reversible reaction for the oxidation of cysteine and indicate how the production of cystine in a protein might stabilize the tertiary structure of proteins.

5. Draw the structure for a tripeptide, pointing out (1) the peptide linkages, (2) the chiral carbon atoms in the backbone, (3) the N- and C-terminal amino acids and (4) the R groups.

6. Name the four levels of structure in proteins and briefly describe what each is associated with.

7. Indicate the four forces which are involved in determining protein structure and draw a

Chapter 21 Proteins

representation of each.

8. Characterize the secondary structures in α-helix and β-pleated sheet in terms of structure and the types of forces involved in each.

9. Indicate what changes occur in proteins by the addition of specific denaturation agents and list some practical applications of denaturation.

10. Define the important terms and comparisons in this chapter and give specific examples where appropriate.

IMPORTANT TERMS AND COMPARISONS

Catalyst
Enzyme
Hydrophobic and Hydrophilic Side Chains
Isoelectric Point, pI
Fibrous and Globular Proteins
R Group or Side Chain
Prion and Pathological Conditions
Quaternary Structure
Triple Helix Structure of Collagen
Prosthetic Group
Enantiomers

Di-, Tri- and Polypeptide
D- and L- Structural Forms
Amphoteric
Buffer Solution
Cysteine and Cystine
Chaperones
O-Linked and N-Linked Saccharides
Glycation and AGE

Protein and Peptide
Amino Acid and Amino Acid Residue
Zwitterion
Hypoglycemia
N-Terminal and C-Terminal Amino Acids
Peptide Backbone
Primary, Secondary and Tertiary Structure
Salt Bridges and Hydrophobic Interactions
Sickle Cell Anemia
Homo- and Heterozygote
Intra- and Intermolecular Hydrogen Bonding

α-Helix, β-Pleated Sheet and Random Coil
Tropocollagen and Collagen
Disulfide Bridge
Hemoglobin
Globin
Glycoproteins
Gelatin

SELF-TEST QUESTIONS

MULTIPLE CHOICE. In the following exercises, select the correct answer from the choices listed. In some cases, two or more choices will be correct.

1. The number of tetrapeptides that can be made from any of the 20 amino acids is:
   a. 1.6×10^5
   b. 20
   c. 4
   d. 80

2. Glycine is an unique amino acid because it:
   a. has no chiral carbon
   b. has a sulfur containing R group
   c. cannot form peptide bond
   d. is an essential amino acid

3. The α-helix is a common form of:
   a. primary structure
   b. secondary structure
   c. tertiary structure
   d. a denatured region

Chapter 21 Proteins 138

4. Which of the following is a transport protein?
   a. hemoglobin
   b. pepsin
   c. collagen
   d. oxytocin

5. A zwitterion has which of the following characteristics?
   a. a high melting point
   b. no net charge
   c. soluble in water
   d. all of above

6. Arginine is a basic amino acid while alanine is nonpolar. The isoelectric point for arginine is:
   a. higher than for alanine
   b. lower than for alanine
   c. about the same as alanine
   d. cannot predict

7. Insulin is a hormone made up of two chains, crosslinked by two disulfide bonds. This

   a. 2 subunits
   b. 1 subunit
   c. no subunits
   d. 4 subunits

8. Which of the following R groups associated with an amino acid can take part in hydrogen bonding?
   a. $-CH_2SCH_3$ (methionine)
   b. $-CH_2OH$ (serine)
   c. $-CH_2COOH$ (aspartic acid)
   d. $-CH_2CH_2CH_2CH_2NH_3^+$ (lysine)

9. Hydrophobic interactions occur between the R groups with:
   a. acidic character
   b. nonpolar character
   c. basic character
   d. all of above

10. The number of amino acid residues in the cyclic nonapeptide, vasopressin, is:
    a. 7
    b. 9
    c. 19
    d. 90

11. An example of a globular protein that is a hormone is
    a. hemoglobin
    b. keratin
    c. collagen
    d. insulin

12. The secondary structure in an alpha helix is stabilized (held together) by
    a. intramolecular hydrogen bonds
    b. hydrophobic interactions
    c. electrostatic bonds
    d. intermolecular hydrogen bonds

13. An example of a prosthetic group is
    a. methionine
    b. the heme group
    c. an amide group
    d. any functional group

COMPLETION. Write the word, phrase or number in the blank space in answering the question.

1. In the tripeptide drawn below, circle and clearly label the following units: peptide linkages; C- and N-terminal ends; stereocenters; R groups.

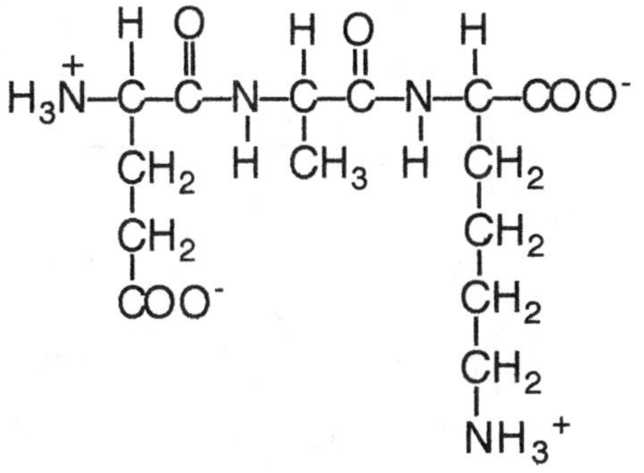

Chapter 21 Proteins

[Structure diagram of a tetrapeptide: H₃N⁺-CH(CH₂CH₂COO⁻)-CO-NH-CH(H)-CO-NH-CH(CH₃)-CO-NH-CH(CH₂CH₂CH₂CH₂NH₃⁺)-COO⁻]

2. Tropocollagen is the _____ form of collagen and is made up of _____ _____ units. Insoluble collagen results from covalent _____ of two _____ residues on adjacent chains in the helix.

3. A covalent bond in the _____ _____, is the force which is associated with the _____ structure of a protein.

4. Agents that can reversibly denature a single subunit protein do this by breaking forces involved in the _____ and _____ structure.

5. Sickle cell trait occurs in people who have _____ gene(s) programmed to produce sickle cell hemoglobin.

6. The α-helix is stabilized by _____-molecular hydrogen bonding, while the β-pleated sheet structure is stabilized by _____-molecular hydrogen bonding. The interaction is in both cases between the _____ and the _____ groups.

7. N-Linked saccharides have the saccharide linked directly to _____, while O-linked saccharides may be linked to either _____, _____ or _____.

8. Complete the following reactions:

a. 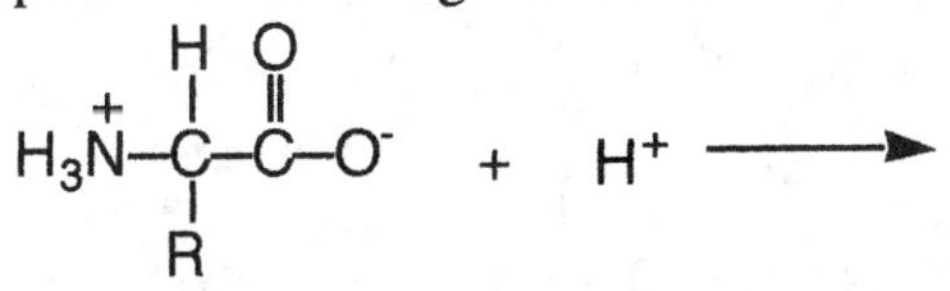 H₃N⁺-CH(R)-C(=O)-O⁻ + H⁺ ⟶ _____

b. 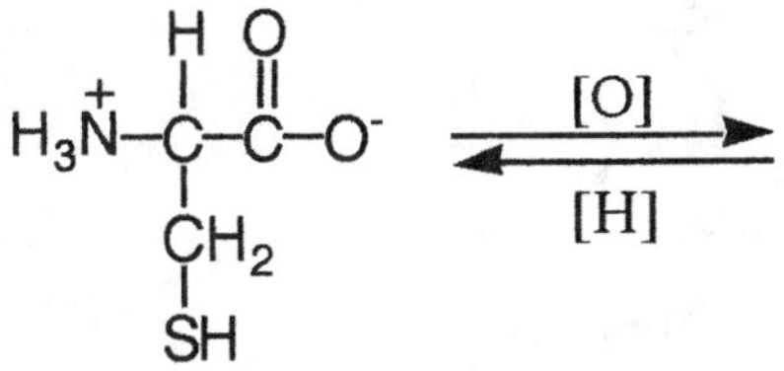 H₃N⁺-CH(CH₂SH)-C(=O)-O⁻ ⇌ [O]/[H] _____

c. tetrapeptide + H₂O →(HCl, heat) _____

Chapter 21 Proteins

d. [structure of alanine + phenylalanine reaction]

alanine + phenylalanine ⟶ _____

9. _____ structure occurs only in proteins that have _____ subunits, such as in hemoglobin.

10. The charge on hydrophobic amino acid, alanine, at very low pH is _____, while at high pH it is _____.

ANSWERS TO SELF-TEST QUESTIONS

Multiple Choice
1. a
2. a
3. b
4. a
5. d
6. a
7. b (Figure 21.3 in text)
8. b, c, d
9. b
10. b
11. d
12. a
13. b

Completion

1.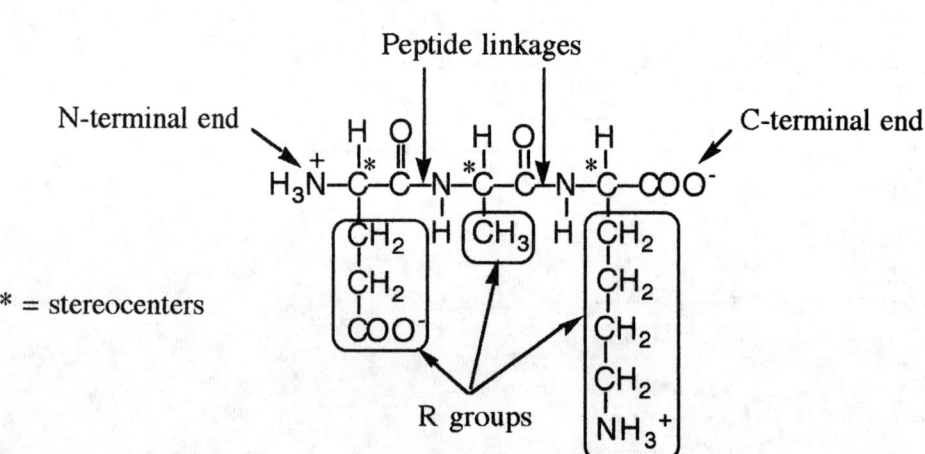

2. soluble; triple helical; crosslinking; lysine
3. peptide linkage; primary
4. secondary; tertiary
5. one

Chapter 21 Proteins 141

6. intra; inter; $\diagdown C=O$; $H-N\diagup$

7. asparagine, serine, threonine, hydroxylysine

8.
   a. 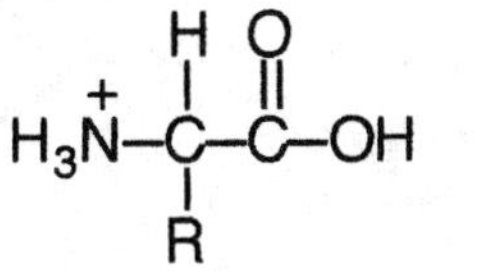 $H_3\overset{+}{N}-\underset{R}{\overset{H}{C}}-\overset{O}{\overset{\|}{C}}-OH$

   b. 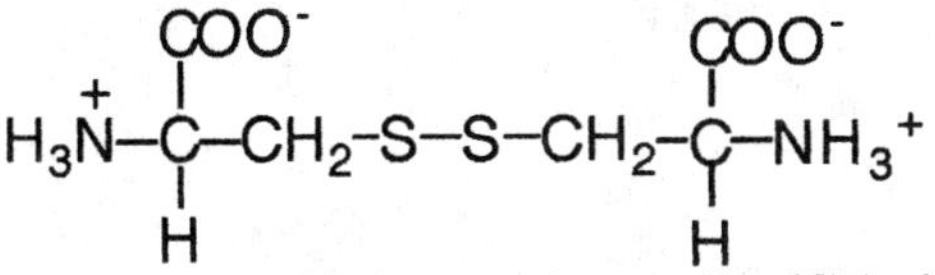 $H_3\overset{+}{N}-\underset{H}{\overset{COO^-}{C}}-CH_2-S-S-CH_2-\underset{H}{\overset{COO^-}{C}}-NH_3^+$

   cystine, which contains a disulfide bond

   c. 4 amino acids

   d. 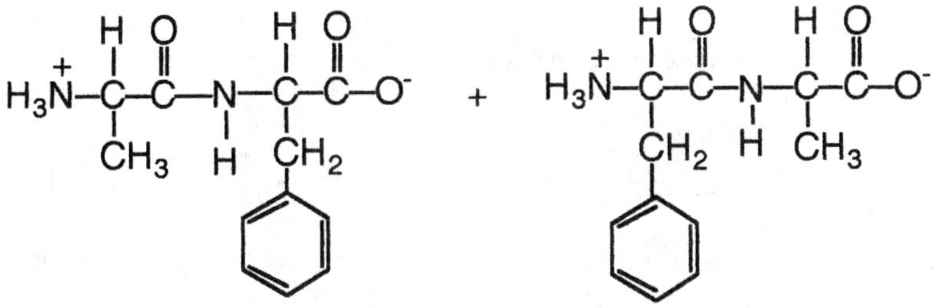 [dipeptide Ala-Phe] + [dipeptide Phe-Ala]

9. Quaternary, multiple (2 or more)
10. +1, -1

Were it not equipped with catalysts, every
living unit would be a static system
- F.G. Hopkins
   Science, 78, 219 (1933)

22

Enzymes

CHAPTER OBJECTIVES

After you have studied the chapter and worked the assigned exercises in the text and the study guide, you should be able to do the following.

1. Explain the statement that enzymes increase the rate of the reaction, but do not change the position of the equilibrium.

2. List the six major categories of enzymes and generally indicate the type of reaction that the enzyme catalyzes.

3. Distinguish between the terms enzyme, cofactor, coenzyme, apoenzyme, proenzyme and active site.

4. Describe, in both words and in figures, the effect the following have on the rate of a catalyzed reaction:
    enzyme concentration
    substrate concentration
    temperature
    pH
    enzyme inhibitors

5. Outline and compare the proposals in the (1) lock and key model and the (2) induced fit model to explain enzymatic action and specificity.

6. Describe how a competitive and a non-competitive inhibitor produce their effect.

7. Define active site and list factors associated with it which are responsible for its specificity toward molecular substrates.

8. Explain how the activity of an allosteric enzyme can be modulated in a positive or negative way by regulator interactions. Relate this effect to the feedback control mechanisms found in

Chapter 22 Enzymes   143

complex reaction systems.

9. Describe how sulfa drugs can be used as a competitive inhibitors to kill harmful bacteria in our bodies, but have no effect on our metabolism.

10. Define the important terms and comparisons in this chapter and give specific examples where appropriate.

IMPORTANT TERMS AND COMPARISONS

Enzyme and Catalyst
-ase
Oxidoreductase
Transferase
Hydrolase
Lyase
Isomerase
Ligase
Cofactor and Coenzyme
Active Site and Regulator Site
Regulator and Regulator Site
Apoenzyme and Holoenzyme
Enzyme Modification
Feedback Control of Enzyme Activity

Substrate and Competitive Inhibitor
Proenzyme and Zymogens
Reversible Inhibitor
Competitive and Non-Competitive Inhibitor
Irreversible Inhibitor
Sulfa Drugs and Para-Aminobenzoic Acid
Negative and Positive Modulation
Enzyme-Substrate Complex
Allosteric Enzyme
Substrate and Active Site
Isozymes
Lock-in Key and Induced-Fit Models
   For Allosteric Enzymes

SELF-TEST QUESTIONS

MULTIPLE CHOICE. In the following exercises, select the correct answer from the choices listed. In some cases, two or more choices will be correct.

1. A molecule which is structurally similar to the substrate for an enzyme will probably be a:
   a. competitive inhibitor
   b. cofactor
   c. regulator
   d. noncompetitive inhibitor

2. The site on an allosteric enzyme that is directly involved in modulation is called
   a. the prosthetic group
   b. regulatory site
   c. active site
   d. target site

3. The regulatory sequence of reactions shown below uses a feedback mechanism of inhibition. The last product, D, in the sequence usually:
   a. inhibits E_1
   b. inhibits E_2
   c. inhibits E_3
   d. inhibits all enzymes

   $$A \xrightarrow{E_1} B \xrightarrow{E_2} C \xrightarrow{E_3} D$$

4. Competitive inhibition can be overcome by:
   a. increasing substrate concentration
   b. increasing pH
   c. decreasing temperature
   d. all of above

Chapter 22 Enzymes

5. The pH influences most enzymatic reactions because pH changes can:
   a. hydrolyze the protein
   b. produce protonation or deprotonation of essential amino acid residues in the active site
   c. change its primary structure
   d. effect the optical activity of the protein

6. A significant increase in the temperature (as when a patient has a very high fever) in an enzymatic reaction will reduce the rate because the:
   a. enzyme acts on itself as an inhibitor
   b. protein is partially or completely denatured
   c. protein undergoes hydrolysis
   d. activation energy is decreased

7. Consider the possible shapes that a molecule might have to act as a competitive inhibitor to the reaction shown below:

   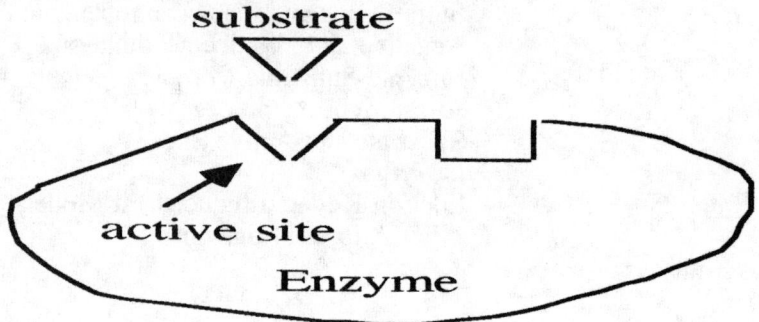

   Indicate which of the following molecules (molecular shapes) could act as a competitive inhibitor.

8. The apoprotein corresponds to which part of an enzyme?
   a. coenzyme   c. cofactor
   b. protein portion   d. part cleaved off in a proenzyme

9. The site on the enzyme at which the substrate interacts is called the
   a. regulatory site   c. active site
   b. modulator site   d. allosteric site

10. Consider the reaction A + B <-----> C + D. An enzyme used in this reaction will:
    a. increase the equilibrium constant
    b. increase the amount of C and D produced
    c. increase the rate at which C and D are produced
    d. lower the activation energy for both the forward and reverse reaction

11. An allosteric enzyme has which of the following characteristics?
    a. A modulator can strongly to the active site

Chapter 22   Enzymes   145

   b.   The binding of a modulator molecule to the regulatory site always inhibits enzymatic activity
   c.   The binding of the modulator to the regulatory site can either increase or decrease the enzymatic activity
   d.   The active site and the regulatory sites are always in different locations and cannot be identical.
   e.   In most cases, the allosteric enzyme is a multisubunit protein.

12. Characteristics of a proenzyme are
   a.   they are made in the cell in an inactive form
   b.   they do not contain all the amino acid residues for a complete and active enzyme
   c.   they are usually cleaved (cut) to produce a smaller protein that becomes the active enzyme
   d.   they do not have the correct prosthetic group for activity

13. The protein modification that is most common in the activation of enzymes is
   a.   methylation
   b.   phosphorylation
   c.   addition of an aromatic group
   d.   addition of an ester group

14. The enzyme, prostaglandin enderoperoxide synthase (PGHS), (Box 22G) is <u>unusual</u> in that it
   a.   has two active sites and catalyzes two different reactions
   b.   has two regulatory sites and no active site
   c.   can be inhibited by common over-the-counter-drugs
   d.   is primarily a polysaccharide

15. The small molecule, urea, is used to denature proteins. Its structure is $H_2N-C-NH_2$
    Its ability to denature proteins is primarily because $\|$
                                                                                        O
   a.   it cleaves the peptide backbone
   b.   it breaks up essential hydrogen bonding in the secondary and tertiary structure
   c.   it breaks up hydrophobic interactions in the secondary and tertiary structure
   d.   it reacts with the hydroxyl group on glycine residues

<u>Completion</u>.   Write the word, phrase or number in the blank space in answering the question.

1. Classify each of the following enzymes into one of the six major classes.

                                                    <u>Class</u>
   a.   Ribonuclease
   b.   Alcohol Dehydrogenase
   c.   DNA Topoisomerase
   d.   Pepsin
   e.   Alanine Racemase
   f.   Pyruvate Kinase
   g.   Acetylcholinesterase
   h.   Aspartate Amino Transferase

2. Lipases are enzymes that are (<u>more or less</u>) specific than trypsin.

Chapter 22 Enzymes

3. Of the two main models used to explain enzymatic action, the _____ model is regarded as more flexible than the _____ model.

4. A number of enzymes require a non-protein component called a _____ or _____ to exhibit enzymatic activity.

5. Trypsinogen is an example of a _____ that can only be _____ by removal of a portion of the protein.

6. If the enzyme concentration as an enzymatic reaction is tripled, the reaction rate will _____.

7. Heavy metals, such as Hg^{2+} or Pb^{2+}, act as _____ inhibitors to many enzymes.

8. Label the following plot to show a pH profile for an enzyme (plot enzyme rate versus pH) in which the highest activity occurs at pH 4.5 and decreases to nearly zero at pH 7.0 and 2.0.

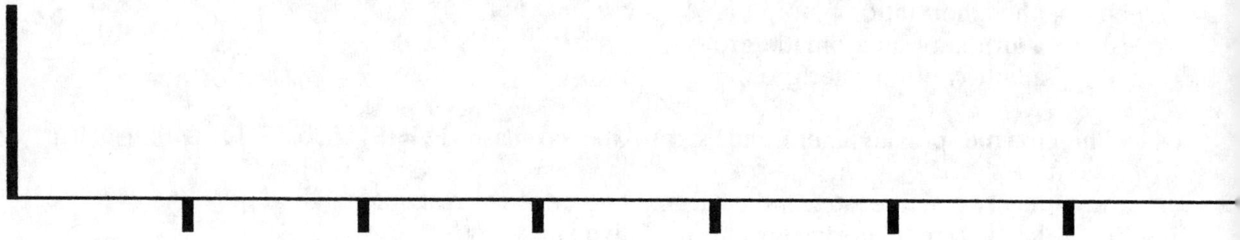

9. Draw the substrate, enzyme and enzyme-substrate (E-S) complex as they may be viewed in the induced fit model.

```
                    -------->
_____   _____   <--------   _____
 Substrate   +   Enzyme                  (E-S) complex
```

10. Although cofactors can be either _____ or _____, coenzymes are always _____ and are synthesized from a vitamin.

ANSWERS TO SELF-TEST QUESTIONS

Multiple Choice
1. a
2. b
3. a
4. a
5. b
6. b
7. a, b (note that the left side of b will inhibit substrate binding)
8. b
9. c
10. c, d
11. c, d, e

Completion

1. a. Hydrolase
   b. Oxidoreductase
   c. Isomerase
   e. Isomerase
   f. Transferase (specific for phosphates)
   g. Hydrolase

Chapter 22 Enzymes

    d. Hydrolase   h. Transferase
2. less
3. induced fit; lock and key
4. cofactor, coenzyme
5. proenzyme, activated
6. triple
7. irreversible
8.

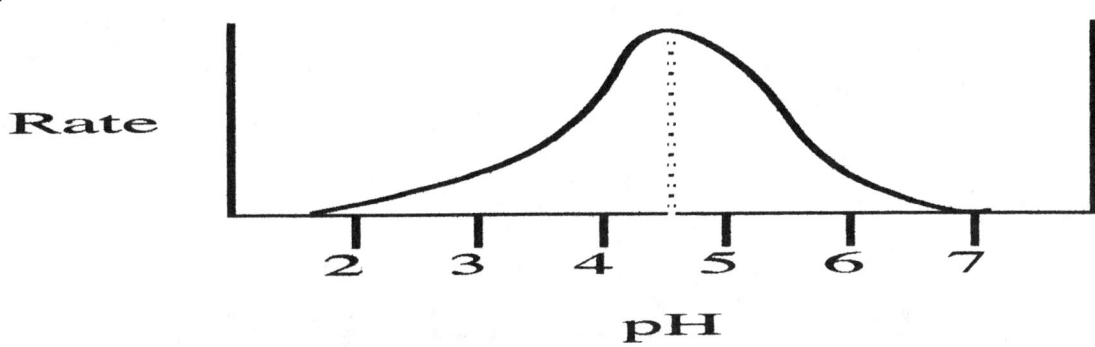

9. The (free) enzyme structure (without substrate bound to it) is different than when the substrate is bound to the active site.

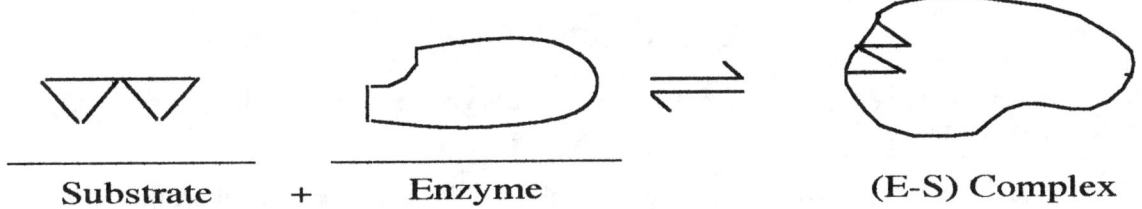

10. inorganic ions (Mg^{2+}, Zn^{2+}, etc). organic compounds, organic compounds

One, if by land, or two, if by sea.
- Paul Revere's Ride
  Henry Wadsworth Longfellow

23

Chemical Communications: Neurotransmitters and Hormones

CHAPTER OBJECTIVES

After you have studied the chapter and worked the assigned exercises in the text and the study guide, you should be able to do the following.

1. Tabulate the two general types of chemicals involved in intercellular communication, indicating (1) their general make up, (2) what each acts on, (3) the time frame of their action, (4) the cells or organs which are involved in the communication and (5) whether secondary messengers are involved in the process.

2. Explain the role of the axons, dendrites, the synapse, vesicles and receptor sites in the neurotransmission by acetylcholine.

3. Describe how acetylcholine bound to postsynaptic receptor sites can be removed normally in the cell and how muscle relaxants or nerve gases can prevent this.

4. Outline the mechanism of action and the final inactivation of adrenergic neurotransmitters by monoamine oxidases (MAO).

5. List the three types of hormones, their general make up and the types of action that each stimulates.

6. List a number of hormones, the gland which secretes the hormone and the specific action or function of the hormone (consult Table 23.2 in text).

7. Define the important terms and comparisons in this chapter and give specific examples where appropriate.

Chapter 23 Chemical Communications

IMPORTANT TERMS AND COMPARISONS

Receptor, Chemical and Secondary Messengers
Agonists and Antagonists
Phosphodiesterase
L-Dopa and Dopamine
Histamine and Histamine Receptors, H_1 and H_2
Cholinergic, Amino Acid, Adrenergic
  And Peptidergic Neurotransmitters
Enkephalins
Acetylcholine
Acetylcholinesterase
Transmembrane Protein and Ion Channels
Primary Messenger
Helicobacter pylori Bacteria and Ulcers
Diabetes mellitus and
  Non-Insulin-Dependent Diabetes
GABA

Adenylate Cyclase and cAMP
Hormone and Endocrine Gland
Axon and Dendrite
Synapse, Presynapse and Postsynapse
Vesicles
Adrenergic, Peptidergic and Steroid
  Hormones
Tamoxifen and Estrogen Receptors
Alzheimer's Dementia
Nerve Gas
Calmodulin
Secondary Messenger
NO and NO Synthase
  G-Protein-cAMP Cascade and
  Signal Transduction

SELF-TEST QUESTIONS

MULTIPLE CHOICE. In the following exercises, select the correct answer from the choices listed.

1. Decamethonium, which has a structure similar to that for acetylcholine, inhibits acetylcholinesterase. It is a
    a. non-competitive inhibitor
    b. chemical mediator
    c. competitive inhibitor
    d. toxic poison

2. Of the natural communication chemicals presented in the chapter, the one which acts over the longer distance?
    a. neurotransmitters
    b. steroid hormones
    c. c-AMP
    d. NO

3. Acetylcholinesterase catalyzes the hydrolysis of which linkage in acetylcholine?

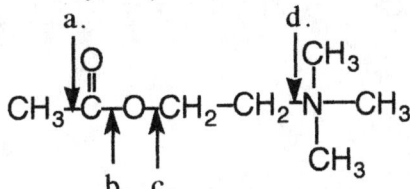

4. H_1 receptors found in the respiratory tract are blocked by the interaction of
    a. insulin
    b. L-DOPA
    b. enkephalins
    d. antihistamines

5. The second messenger in the action of monoamine neurotransmitters is in most cases
    a. d-AMP
    b. c-AMP
    c. AMP
    d. ATP

Chapter 23   Chemical Communications   150

6. The administration of L-Dopa can reverse the symptoms of Parkinson's disease. Dopamine is produced from L-Dopa by a
    a. hydrolysis reaction
    b. oxidative decarboxylation reaction
    c. decarboxylation reactions
    d. oxidative reaction

7. Protein kinases function as
    a. phosphate transfer enzymes
    b. structural membrane proteins
    c. secondary messengers
    d. ion translocating proteins

8. Neurotransmitters are stored in vesicles in the
    a. axon
    b. dendrites
    c. presynaptic site
    d. postsynaptic site

9. Glucagon is an example of a
    a. peptide hormone
    b. steroid hormone
    c. monoamine neurotransmitter
    d. receptor

10. An agonist is a molecule which
    a. stimulates antibody production
    b. competitively inhibits acetylcholinesterase
    c. activates receptor sites
    d. reduces pain

11. Steroid hormones produce their effect in the cell by
    a. binding to the membrane
    b. activating a protein kinase
    c. stimulating c-AMP production
    d. affecting the transcription of genes

12. The classes of hormones include all the molecules listed, with the exception of
    a. peptides
    b. nucleotides
    c. steroids
    d. small molecules, such as amino acids

COMPLETION. Write the word, phrase or number in the blank space in answering the question.

1. Complete the following reactions, writing structural formulas for products and/or reactants.

    a. $(CH_3)_3NCH_2CH_2OCOCH_3 + H_2O \rightleftharpoons$ _____ + _____

    b. [ATP structure] + $H_2O \rightleftharpoons$ [diphosphate structure] + _____

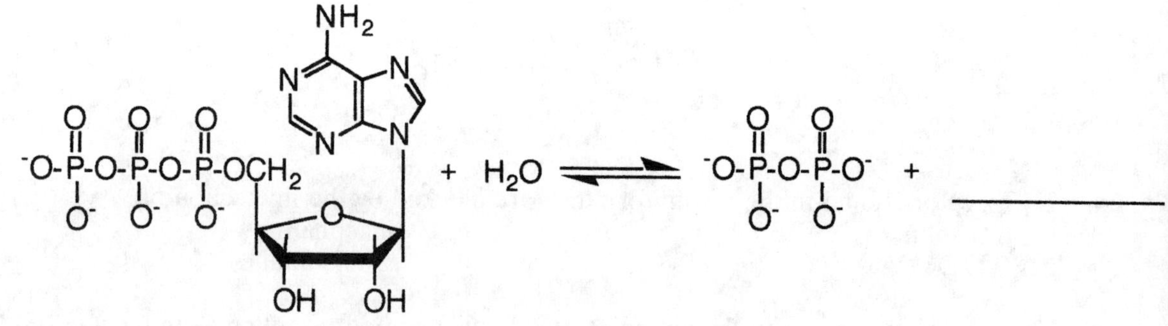

    c. _____ + $HOCH_2CH_2\overset{+}{N}(CH_3)_3 \rightleftharpoons$ _____ + _____

Chapter 23 Chemical Communications   151

2. The _____ is the fluid-filled space between the axon of one neuron and the _____ of the next neuron.

3. A neurotransmitter and a receptor combine very specifically together. This interaction has been likened to the _____ _____ _____ model for enzyme-substrate interactions.

4. Food poisoning from botulism toxin deprives cholinergic nerves from their neurotransmitters. This indicates that the synthesis of _____ is blocked.

5. To degrade the neurotransmitter, acetylcholine, in the postsynaptic receptor site, the enzyme, _____ , acts on acetylcholine to _____ it to acetic acid and choline.

6. Monoamine neurotransmitters, such as norepinephrine, form a membrane bound _____ - _____ like complex, which activates the enzyme _____ that produces the secondary messenger _____ from ATP.

7. GABA is the abbreviation for _____ .

8. The _____ end of methionine enkephalin in similar to a segment of the alkaloid painkiller, _____ .

9. Antidiabetic drugs are useful for diabetic patients who have insufficient numbers of _____ _____ on their target cells. reaction.

10. MAO is the abbreviation for _____, which has the function of _____ monoamine neurotransmitters by _____ them to an _____ functional group.

11. The reaction type that converts histidine to histamine is the same type that converts L-Dopa to dopamine. This type of reaction is called a _____ reaction.

12. Match the disease or condition in the left-hand column with the term in the right-hand column that is most closely associated with the disease or condition.

   a. Alzheimer's disease
   b. Parkinson's disease
   c. Impotency
   d. Diabetes
   e. Botulism
   f. Breast cancer

   i. low insulin levels
   ii. tamoxifen binding to estrogen receptors
   iii. acetylcholine transferase activity
   iv. L-Dopa and inhibitors of monoamine oxidase
   v. inhibitors of phosphodiesterase activity
   vi. Prevention of acetylcholine release from presynaptic vesicles

ANSWERS TO SELF-TEST QUESTIONS

Multiple Choice
1. c
2. a
3. c
4. d
5. b
6. c

7. a
8. c
9. a
10. c
11. d
12. b

Chapter 23 Chemical Communications   152

Completion
1. a. acetylcholinesterase; CH₃COOH + choline
    b. adenylate cyclase; c-AMP (see figure in text)
    c. CH₃C(=O)-SCoA; acetylcholine transferase; CH₃COCH₂CH₂N(CH₃)₃⁺ + CoA-SH
        (with C=O on the COCH₂ group)

2. synapse; dendrites (or cell body)
3. lock-and-key
4. acetylcholine
5. acetylcholinesterase; hydrolyze
6. hormone-receptor; adenylate cyclase; c-AMP
7. γ-aminobutyric acid
8. N-terminal end; morphine
9. insulin receptors
10. monoamine oxidase, inactivating, oxidizing, aldehyde
11. decarboxylation
12. a. iii
    b. iv
    c. v
    d. i
    e. vi
    f. ii

Chapter 24   Nucleic Acids and Heredity

What distinguishes a butterfly from
a lion, a hen from a fly, or a worm
from a whale is........
    -F. Jacob
    Science, 196, 1161 (1977)

24

Nucleotides, Nucleic Acids, and Heredity

CHAPTER OBJECTIVES

    Each one of us is unique in many ways. These characteristics result because each individual has a different genetic makeup in that, although our DNA appears much the same as everyone else's, the sequence (i. e., the order of the bases) of nucleotides in the DNA strands is quite different. These differences become manifest as the genes in the DNA are transcribed into RNA and then translated into the many diverse enzymes and proteins (the real "workers" in the cell), which reveal important aspects of our individuality. In this chapter, we learn about the two types of nucleic acids - DNA and RNA. After first dissecting nucleic acids into their constituents, the general structure of these polynucleotides will be examined. The process of replication by which DNA is duplicated with high fidelity is presented, together will aspects of cloning and a revolutionary technique called Polymerization Chain Reaction (PCR). These concepts will provide the basis for examining how selected genes are expressed in cells (Chapter 25) and lead to an understanding of the Central Dogma of Molecular Genetics. The "flow" of genetic information is briefly summarized on the following page and should be consulted during your study of both Chapters 24 and 25..

Chapter 24 Nucleic Acids and Heredity

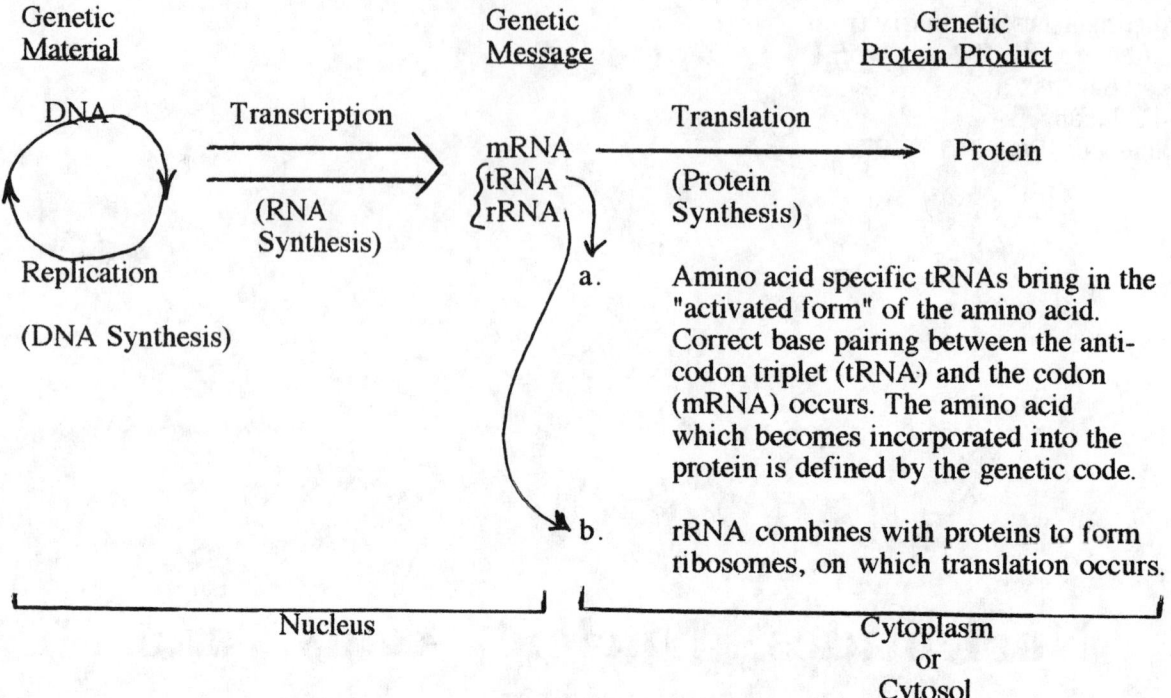

The complexity of DNA is further revealed in the exon and intron segments of the genes in higher organisms and the repeating nucleosome subunits in the nuclear chromosome. These properties were only realized in the mid 1970's.

After you have studied the chapter and worked the assigned exercises in the text and the study guide, you should be able to do the following.

1. Explain the importance of hereditary information and its storage and expression in DNA.

2. Name and draw the structures of the organic bases (purines and pyrimidines), the nucleosides and the nucleotides that occur in DNA and RNA.

3. Draw a dinucleotide unit of DNA containing one purine and one pyrimidine base; point out the 5'- and 3'-ends, the glycosidic bond between the sugar and base and the phosphodiester backbone in this unit.

4. Draw the complementary base pairs, G≡C, and A=T, showing all atoms and the hydrogen bonding interactions.

5. In double-stranded DNA, indicate the relative positions of the aromatic bases, the sugar and phosphate units in this structure.

6. List the compositional, structural and functional differences between DNA and RNA.

7. Outline the main features of DNA replication, indicating the role or significance of unwinding proteins, DNA polymerase, complementary base pairing, semi-conservative replication, 5'---->3' direction, Okazaki fragments and DNA ligase.

8. Describe the structure of a tRNA molecule, the position of the anticodon loop, the role of the 3'-end and the function of this RNA in the translation process.

Chapter 24 Nucleic Acids and Heredity

9. Outline how the PCR technique can amplify DNA and indicate the role of temperature, DNA polymerase and primers play in the procedure.

10. Discuss the role of telomerase in DNA replication and its proposed role in immortality.

11. Define the terms gene, exons and introns as they relates to higher organisms.

12. Describe the nucleosome unit in terms of its composition and structure.

13. Explain the immediate goal of the Human Genome Project and its importance in the understanding of human biology.

14. Indicate the important new use of DNA fingerprinting in forensic science.

15. Define how apoptosis differs from necrosis and indicate the role of caspases and endonuclease in the apoptotic process.

16. Define the important terms and comparisons in this chapter and give specific examples where appropriate.

IMPORTANT TERMS AND COMPARISONS

Zygote
Nucleus (of the cell)
Ribonucleotides and Deoxyribonucleotides
Histone Protein

Nucleic Acid
Deoxyribonucleic Acid (DNA)
Ribonucleic Acid (RNA)
Purine Bases; A, G
Anticancer Drug and Chemotherapy
3'- and 5'-Carbon on Sugar
(Inheritable) Gene
Genes and Exons and Introns

A=T and G≡C Hydrogen Bonding
DNA Polymerase
Replication Fork
Leading and Lagging DNA Strands
Helicases
Satellite DNA
Clover Leaf Structure of tRNA
Restriction Enzyme
mRNA, rRNA, tRNA and Ribozymes
DNA Cloning
Heat-Resistant DNA Polymerase
Apoptosis
Somatic and Germ Line Cells

Central Dogma of Molecular Biology
Chromosome
Transcription and Translation
Nucleosomes and Histone Proteins

β-N-Glycosidic Bond
One Gene-One Enzyme Hypothesis
Pyrimidine Bases; C, T, U
D-Ribose and D-Deoxyribose
Nucleoside and Nucleotide
Primary and Secondary Structure of DNA
3'-and 5'-OH Ends of DNA
Complementary Base Pairs

Semiconservative DNA Replication
Okazaki Fragments
Escherichia coli (E. coli)
DNA Ligase
Continuous and Discontinuous
    Strand Synthesis
Splicing of mRNA
Telomeres and Telomerase
Polymerase Chain Reaction (PCR)
DNA Fingerprinting
Human Genome Project
Caspases

Chapter 24 Nucleic Acids and Heredity 156

SELF-TEST QUESTIONS

MULTIPLE CHOICE. In the following exercises, select the correct answer from the choices listed. In some cases, two or more choices will be correct.

1. The backbone in a nucleic acid is called a
   a. glycosidic bond
   b. sugar backbone
   c. phosphodiester backbone
   d. peptide backbone

2. The coding sequences of a gene in DNA are called
   a. nucleosomes
   b. introns
   c. exons
   d. initiating factors

3. An linkage connecting the base and the deoxyribose is a
   a. peptide backbone
   b. β-N-glycosidic bond
   c. phosphodiester backbone
   d. other linkage

4. A nucleosome is made up of
   a. DNA and histone proteins
   b. rRNA and proteins
   c. nucleosides
   d. both DNA and RNA

5. Telomerase is an enzyme that
   a. is made of both an RNA and protein subunit
   b. replicates the telomers at the ends of the chromosomes
   c. replicates DNA in a semiconservative manner
   d. can be used in PCR

6. A feature which distinguishes ribose from deoxyribose is the absence of an OH at the
   a. 2'-carbon
   b. 3'-carbon
   c. 1'-carbon
   d. 5'-carbon

7. The hydrogen bonding between a G and C in forming a GC base pair in DNA involves
   a. atoms on or in the six-membered rings of both guanine and cytosine bases
   b. atoms on or in the six-membered-ring of cytosine and the five-membered ring in guanine
   c. only oxygen atoms in cytosine and nitrogen atoms in guanine
   d. only hydrogen-bond donors in cytosine and only hydrogen-bond acceptors in guanine

8. Adenine is the name of a
   a. nucleoside
   b. purine base
   c. nucleotide
   d. amino acid

9. The sequence GCCCTGA has the A at the
   a. 5'-end
   b. 3'-end
   c. can be either the 5'- or 3'- end
   d. depends on size of DNA

10. The 3'-end of a tRNA molecule
    a. contains the anticodon loop
    b. binds to a specific amino acid
    c. has no known function
    d. binds to rRNA

11. The structure of cytosine is

a.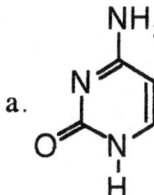

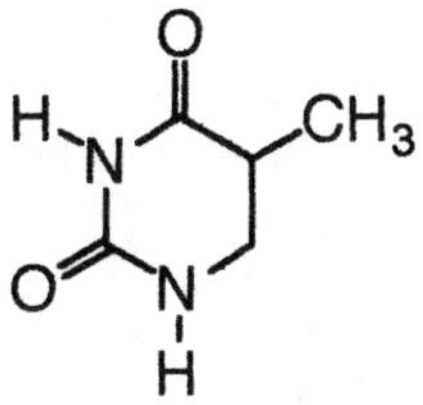

b.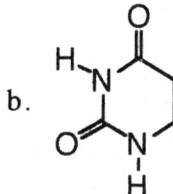

c. [structure shown]

d. 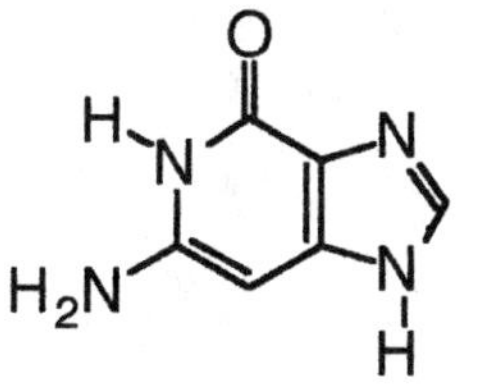 [structure shown]

12. DNA fingerprinting requires the use of
    a. protein synthesis
    b. transcription factors
    c. restriction enzymes
    d. exons and introns

13. Okazaki fragments are
    a. formed in RNA transcription
    b. formed in DNA replication
    c. small RNA molecules
    d. small DNA fragments of about 1000 nucleotides each

14. A glycosidic bond in a nucleoside or nucleotide links the (deoxy)ribose sugar and which position on a purine base?
    a. N-1 atom
    b. N-9 atom
    c. N-3 atom
    d. C-8 atom

15. One of the complementary base pairs in DNA is
    a. AU
    b. GC
    c. AT
    d. GT

16. An RNA molecule has no
    a. uracil bases
    b. thymine bases
    c. cytosine bases
    d. guanine bases

17. The hydrogen bonding in an AT base pair involves atoms in the purine ring exclusively associated with the
    a. 6-membered ring
    b. 5-membered ring
    c. both the 5- and 6-membered rings
    d. none of the above

18. The process by which cells become specialized (neurons, muscle, etc.) in the development of a multicellular organism is called
    a. replication
    b. differentiation
    c. cell division
    d. genetic mutation

Chapter 24 Nucleic Acids and Heredity   158

19. The single-stranded DNA primers used in PCR can be complementary to
    a. only the 3-ends of the DNA to be amplified
    b. only the 5'-end of the DNA to be amplified
    c. either the 3- or the 5-ends of the DNA to be amplified
    d. both the 3'- or the 5-ends of the DNA

20 Caspases are
    a. nucleases that cut at cytidine residues in DNA
    b. enzymes used in cloning of DNA
    c. enzymes that cleave proteins next to an aspartic acid residue
    d. involved in programmed cell death called apoptosis

COMPLETION. Write the word, phrase or number in the blank space in answering the question.

1. A DNA molecule is thought to have as many as a _____ genes.

2. Replication of DNA in eukaryotic cells occurs in the _____ of the cell, while translation occurs in the _____.

3. There are _____ hydrogen bonds between adenosine and the complementary base, _____.

4. The only nucleic acid that exhibits any enzymatic activity is called a _____ .

5. A _____ is produced on the joining of an egg cell with a sperm cell.

6. Amino acids are attached to the _____ end of tRNA on the _____ unit.

7. Replication of the new DNA strand occurs in a _____ direction, while the DNA polymerase is traveling in the _____ direction on the DNA template strand.

8. _____ are composed of rRNA molecules and proteins and are the units on which protein synthesis occurs in the _____ of the cell.

9. The organic bases in DNA reside on the interior of the double helix and stabilize the structure by _____ interactions.

10. _____ bases are made up of two fused rings, while _____ bases have only a six-membered heterocyclic ring.

11. The sequence of one strand of double stranded DNA is (5')AGGAGCTTCG(3'). The sequence of the complementary strand is _____.

Chapter 24 Nucleic Acids and Heredity

12. Match the scientist(s) in the left-hand column with the accomplishment on the right with which he (they) is most closely associated.
    a. Okazaki
    b. J. Watson & F. Crick
    c. K. Mullis
    d. O. Avery
    e. R. Franklin & M. Wilkins
    f. G. Beadle & E. Tatum
    g. E. Chargaff

    (i)   Postulated structure of DNA
    (ii) Proposed DNA contains the genetic information
    (iii) X-ray studies of DNA
    (iv) Proposed one gene-one enzyme hypothesis
    (v)   Discovered that in DNA,
          no. of A = no. of T
          no. of G = no. of C
    (vi) Discovered PCR technique
    (vii) Discovered small segments of DNA made in DNA replication

ANSWERS TO SELF-TEST QUESTIONS

Multiple Choice
1. c
2. c
3. b
4. a
5. a, b
6. a
7. a
8. b
9. b
10. b
11. a
12. c
13. b
14. b
15. b, c
16. b
17. a
18. b
19. a
20. c, d

Completion
1. 100,000
2. nucleus; cytoplasm (or cytosol)
3. two; thymine
4. ribozyme (or catalytic RNA)
5. zygote
6. 3'; adenosine
7. 5'-->3', 3'-->5'
8. Ribosomes, cytoplasm
9. hydrophobic
10. Purine; pyrimidine
11. (3')-TCCTCGAAGC-(5')
12. a. (vii)
    b. (i)
    c. (vi)
    d. (ii)
    e. (iii)
    f. (iv)
    g. (v)

Chapter 25 Transfer of Information: Protein Synthesis

It is the amazement of selective gene expression
that destines certain cells to become liver cells,
others become heart cells, etc.- all remarkably orchestrated
to make you, me and our neighbor.
  -Just thoughts

25

Gene Expression and Protein Synthesis

Each one of us is unique. However, the liver in every one of us functions in the same way, likewise the heart, the colon, etc. These characteristics result from the finding that although every cell contains the same genetic material, DNA, with its many genes, only a small select collection of these genes is expressed into proteins in any particular cell type (tissue). This chapter permits us to begin to understand these remarkable happenings within us.

CHAPTER OBJECTIVES

After you have studied the chapter and worked the assigned exercises in the text and the study guide, you should be able to do the following.

1. Consider a gene and go through the details of the step-wise processes of transcription and translation

2. Outline the transcription process and the role of transcription factors, mRNA and the codon units in the transfer of genetic information.

3. Describe the structure of a tRNA molecule, the position of the anticodon loop, the role of the 3'-end and the function of this RNA in the translation process.

4. Describe the location and role of rRNA in the translation process.

5. Explain what the genetic code is and the importance of the genetic code in converting mRNA into protein.

6. Describe the processes of amino acid activation and then the initiation, elongation and termination steps in protein synthesis.

7. Describe the role of the A and P sites and the 40S and 60S ribosomes in translation.

8. Define the terms, gene, exons and introns as they relate to higher organisms.

9. Explain how a mutation can be produced during DNA replication or by chemicals called mutagens.

10. Explain the use of plasmids, restriction endonucleases and Escherichia coli in recombinant DNA technology.

11. Describe the role of nucleoside-analog-inhibitors and protease inhibitors in the treatment of patients with AIDS.

12. Describe the role of the promoter and the operator sites in the regulation of the expression of structural genes in an operon.

13. Explain the difference between a proto-oncogene and an oncogene and indicate its importance in the life of the cell.

14. Define the important terms and comparisons in this chapter and give specific examples where appropriate.

IMPORTANT TERMS AND COMPARISONS

Central Dogma of Molecular Biology
Transcription, Translation and Gene Expression
Ribonucleic Acid (RNA)
One Gene-One Enzyme Hypothesis
Transcription Factors
Codon (Triplet Sequence) in mRNA
Universal Genetic Code
Degenerate or Multiple Code
40S and 60S Ribosome Particles
Activated Amino Acid Synthetase
40 S and 60 S Ribosomes
Antibiotics and Antiviral Agents
Nucleoside-Analog-Inhibitors
Protease Inhibitors for AIDS Treatment
Operon
Promoter and Operator Sites
Lac Operon for Lactose Metabolism
Transcription Regulatory Factors
  (That Bind Metal Ions (Zn^{+2})
Nucleotide Excision Repair (NER)
Tumor Suppressor Gene and p53
Recombinant DNA Techniques
Restriction Endonuclease
DNA Ligase
Genetic Engineering

Nucleus (of the Cell)
RNA Polymerase
mRNA, rRNA and tRNA
Initiation and Termination Signals
Svedberg Unit, S
Anticodon (Triplet Sequence) in tRNA
Initiation and Termination Codons
Clover Leaf Structure for tRNA
Activated Amino Acid
Amino Acid-tRNA
P and S Sites on 60S Ribosome
Retrovirus and the AIDS Virus
AZT, ddC, ddI and 3TC
Escherichia coli Bacterium
Regulatory Sequences and Structural
  Genes
Response Sequences in Promoters
Mutagens and Mutations

Mutagens and Carcinogens
Proto-oncogenes and Oncogenes
Inborn Genetic Disease
Plasmids
"Sticky Ends"
Gene Therapy
(Inheritable) Gene

Chapter 25 Transfer of Information: Protein Synthesis

SELF-TEST QUESTIONS

MULTIPLE CHOICE. In the following exercises, select the correct answer from the choices listed. In some cases, two or more choices will be correct.

1. The central dogma of molecular biology states that
   a. DNA is a double stranded helix
   b. Information flows from DNA to RNA to proteins
   c. Gene expression is different in all tissues
   d. DNA replication is the key event in cell division

2. The coding sequences of a gene in DNA are called
   a. nucleosomes
   b. introns
   c. exons
   d. initiating factors

3. An RNA virus contains which of the following
   a. DNA
   b. RNA
   c. proteins
   d. carbohydrates

4. The RNA molecules that are translated into proteins are
   a. tRNA
   b. rRNA
   c. mRNA
   d. ribozymes

5. The three-letter nucleotide sequence that is listed for the genetic code is the sequence that is found on
   a. codon sequence of mRNA
   b. sequence on DNA
   c. anti-codon sequence of tRNA
   d. codon sequence on rRNA

6. In protein synthesis, the initial amino acid is always
   a. glycine
   b. cysteine
   c. methionine
   d. alanine

7. Translation occurs
   a. on the ribosomes in the cytoplasm
   b. with the first amino acid being the N-terminal residues
   c. on the ribosomes in the nucleus
   d. with the first amino acid being the C-terminal residue

8. The activation of the amino acid prior to translation requires the consumption of how many high-energy phosphate bonds?
   a. none
   b. one
   c. two
   d. three

9. The 3'-end of a tRNA molecule
   a. contains the anticodon loop
   b. binds to a specific amino acid
   c. has no known function
   d. binds to rRNA

10. The bond that links the amino acid residue to AMP is an
    a. ether linkage
    b. acid anhydride linkage
    c. ester linkage
    d. amide linkage

Chapter 25 Transfer of Information: Protein Synthesis

11. The initiating codon in protein synthesis is
    a. GGG
    b. GUA
    c. AUG
    d. ATG

12. Antibiotics are used to
    a. kill viruses
    b. inhibit DNA replication
    c. bind to mutagens
    d. kill bacteria

13. The ribosome unit that contains the P and A sites is
    a. the 60S ribosome
    b. the 40S ribosome
    c. both the 40S and 60 S
    d. on the tRNA

14. The first amino acid residue incorporated into a protein is always a
    a. alanine
    b. glycine
    c. lysine
    d. methionine

15. The important allosteric regulatory protein in the lac operon is
    a. repressor protein
    b. activator protein
    c. RNA polymerase
    d. β-galactosidase

16. In protein synthesis, the mRNA attaches to the
    a. 30 S ribosome
    b. 40 S ribosome
    c. 100 S nucleosome
    d. 60 S ribosome

17. An RNA molecule contains no
    a. uracil bases
    b. thymine bases
    c. cytosine bases
    d. guanine bases

18. Transcription factors bind to the
    a. promoter site
    b. structural gene
    c. operator site
    d. ribosome binding site

19. An amino acid is activated by reaction with
    a. tRNA
    b. synthetase
    c. ATP
    d. methionine

20. Restriction endonucleases have the property that they
    a. can ligate DNA molecules
    b. are required for termination of transcription
    c. cleave DNA in a sequence-specific manner
    d. bind to regulatory sites in promoters

21. Proteins that contain metal binding sites often bind to
    a. Zn^{2+}
    b. Na^+
    c. H^+
    d. Fe^{2+}

22. The operon is a segment of DNA that collectively includes
    a. collection of structural genes
    b. control sites
    c. regulatory genes
    d. all of the above

Chapter 25 Transfer of Information: Protein Synthesis   164

23. A tumor suppressor gene that is mutated in about 40% of all human cancers is
   a. EGF
   b. p53
   c. p100
   d. β-galactosidase

24. A repair mechanism that repairs some mutations is called
   a. nuclear repair machinery
   b. DNA repairase
   c. nucleotide excision repair
   d. mutationase

25. A number of transcription factors have been discovered which are identified by the presence of the following features in these proteins:
   a. leucine zipper
   b. helix-turn-helix motifs
   c. metal-binding-fingers
   d. all of these features

COMPLETION. Write the word, phrase or number in the blank space in answering the question.

1. The _____ site on the ribosome is the site at which the incoming tRNA binds, while the _____ site is the site where the tRNA binds that contains the growing peptide chain.

2. A mutation by a chemical or by radiation occurs directly on _____, but the result is then transmitted to both the intermediate _____ and the final _____ product.

3. Plasmids are small, _____ DNAs found in _____ cells. They are useful in the _____ of genes.

4. Human insulin is one of the first proteins to be produced by _____ _____ _____ (three words).

5. It has been found that all _____ are mutatgenic, but not all mutagens are _____.

6. All amino acids are attached to the _____ end of tRNA on the _____ unit.

7. Referring to the table for the Genetic Code in the text, write the amino acid sequence for the protein produced from the mRNA shown below:
   AUGUUGCACAACGGGGCGGUGGUGUAA

8. After combining the DNA of a plasmid that has sticky ends with another fragment of DNA with the same sticky ends, the enzyme _____ _____ is added to covalently splice the two DNA fragments together.

9. Ionizing radiation, gamma rays and chemicals, including benzene and vinyl chloride, are known to be _____.

10. The sequence of one strand of double stranded DNA is (5')AGGAGCTTCG(3'). The sequence of the RNA strand produced from transcription on this strand is _____.

11. Synthesis of RNA (transcription) occurs in a _____ direction and along the complementary DNA strand in a _____ direction.

12. The development of cancer may occur when a normal gene, called an _____, is altered or _____. The resulting mutated gene is referred to as an _____.

13. Match the scientist(s) or term in the left-hand column with the most closely associated term in the right-hand column.
    a. AIDS
    b. A stop codon
    c. AIDS treatment
    d. p53

    e. Operon
    f. G. Beadle & E. Tatum
    g. M. Nirenberg

    (i) UAA
    (ii) Determined the genetic code
    (iii) Caused by a retrovirus infection
    (iv) Contains regulatory and structural genes
    (v) Proposed one gene-one enzyme hypothesis
    (vi) Most commonly mutated gene in human cancers
    (vii) AZT

ANSWERS TO SELF-TEST QUESTIONS

Multiple Choice
1. b
2. c
3. b, c
4. c
5. a
6. c
7. a, b
8. c
9. b
10. b
11. c
12. d
13. a
14. d
15. a
16. b
17. b
18. a
19. c
20. c
21. a
22. d
23. b
24. c
25. d

Completion
1. A, P
2. DNA, RNA, protein
3. circular, bacterial (E. coli), cloning
4. recombinant DNA technology
5. carcinogens, carcinogenic
6. 3'; adenosine
7. (N-terminal) Met-Leu-His-Asn-Gly-Ala-Val-Val (C-terminal)
8. DNA ligase
9. mutagenic and carcinogenic
10. (3')TCCTCGAAGC(5')
11. 5'------>3'; 3'------>5'
12. proto-oncogene; mutated; oncogene
13. a. iii
    b. i
    c. vii
    d. vi
    e. iv
    f. v
    g. ii

"Whatever a cell does has to be paid
for in the currency of energy"
    - Szent-Gyorgyi

26

Bioenergetics: How the Body Converts Food to Energy

CHAPTER OBJECTIVES

After you have studied the chapter and worked the assigned exercises in the text and the study guide, you should be able to do the following.

1. List the major organelles in a typical animal cell and indicate at least one important biological process that is carried out in each organelle.

2. List the (molecular) components contained in ATP, the coenzymes NAD^+ and FAD, and acetylCoA. Write out a general reaction in which each takes part and draw the part of each structure which is changed as a result of the reaction.

3. Distinguish generally between the catabolic and the anabolic process in terms of the consumption or production of ATP, NAD^+ and FAD.

4. Draw a clear schematic representation of a mitochondrion pointing out the outer membrane, inner membrane, intermembrane region, the matrix, cristae, and the location of the proton translocating ATP-ase.

5. State the major role of ATP in metabolic reactions.

6. Explain the role of NAD^+ and FAD in the citric acid cycle and in the electron transport chain.

7. Describe the role of acetyl CoA in the Krebs cycle.

8. Name the major species in the Krebs cycle.

9. Indicate the type of reaction involved at each step in the Krebs cycle and indicate (in words) an example of each.

10. Explain how the reduced coenzymes, NADH and $FADH_2$, are involved in oxidative phosphorylation and why each produces a different number of ATP molecules.

Chapter 26 Bioenergetics: How the Body Converts Food to Energy

11. Determine the number of ATP molecules produced in the Krebs cycle and oxidative phosphorylation as a result of the oxidation of one acetyl group.

12. Describe the important features of the chemiosmotic hypothesis.

13. Describe the function of the proton translocating ATP-ase.

14. Explain the value of using an uncoupling agent in oxidative phosphorylation studies.

15. Give specific examples of the use of ATP to produce mechanical, electrical and chemical energy.

16. Describe the roles of the enzymes, cytochrome P-450, superoxide dismutase and catalase in specific oxidative reactions.

17. Define the important terms and comparisons in this chapter and give specific examples where appropriate.

IMPORTANT TERMS AND COMPARISONS

Catabolism and Anabolism
Activated C_2 Fragments or Acetyl Coenzyme A
Organelles
Nicotinamide
Niacin and Niacinamide
Mitochondrion-Outer and Inner Membrane, Intermembrane Space, Matrix and Cristae
Superoxide (Anion), O_2^-
Catalase
Krebs Cycle, Tricarboxylic Acid Cycle and Citric Acid Cycle
AMP, ADP, ATP
Adenosine and Adenine
Ribose
Phosphoric Anhydride Linkage
Phosphate Ester Linkage
High Energy Phosphate Bond
Inorganic Phosphate
Turnover Rate
Coenzyme
NAD^+ and NADH
FAD and $FADH_2$

Riboflavin and Flavin
Nicotinic Acid
Nucleus, Lysosomes, Golgi Bodies, Mitochondria
Pantothenic Acid
Mercaptoethylamine
Cytochrome P-450
Superoxide Dismutase (SOD)
Glutathione
Oxaloacetate
Dehydration Reaction
Tertiary and Secondary Alcohol
Oxidative Decarboxylation
Guanine
GTP
Flavoprotein
Q Enzyme
Cytochromes
Proton Translocating ATP-ase
Uncoupling Agent
Protonphore
Myosin and Actin

Chapter 26 Bioenergetics: How the Body Converts Food to Energy

FOCUSED REVIEW

I. Stages of Metabolism

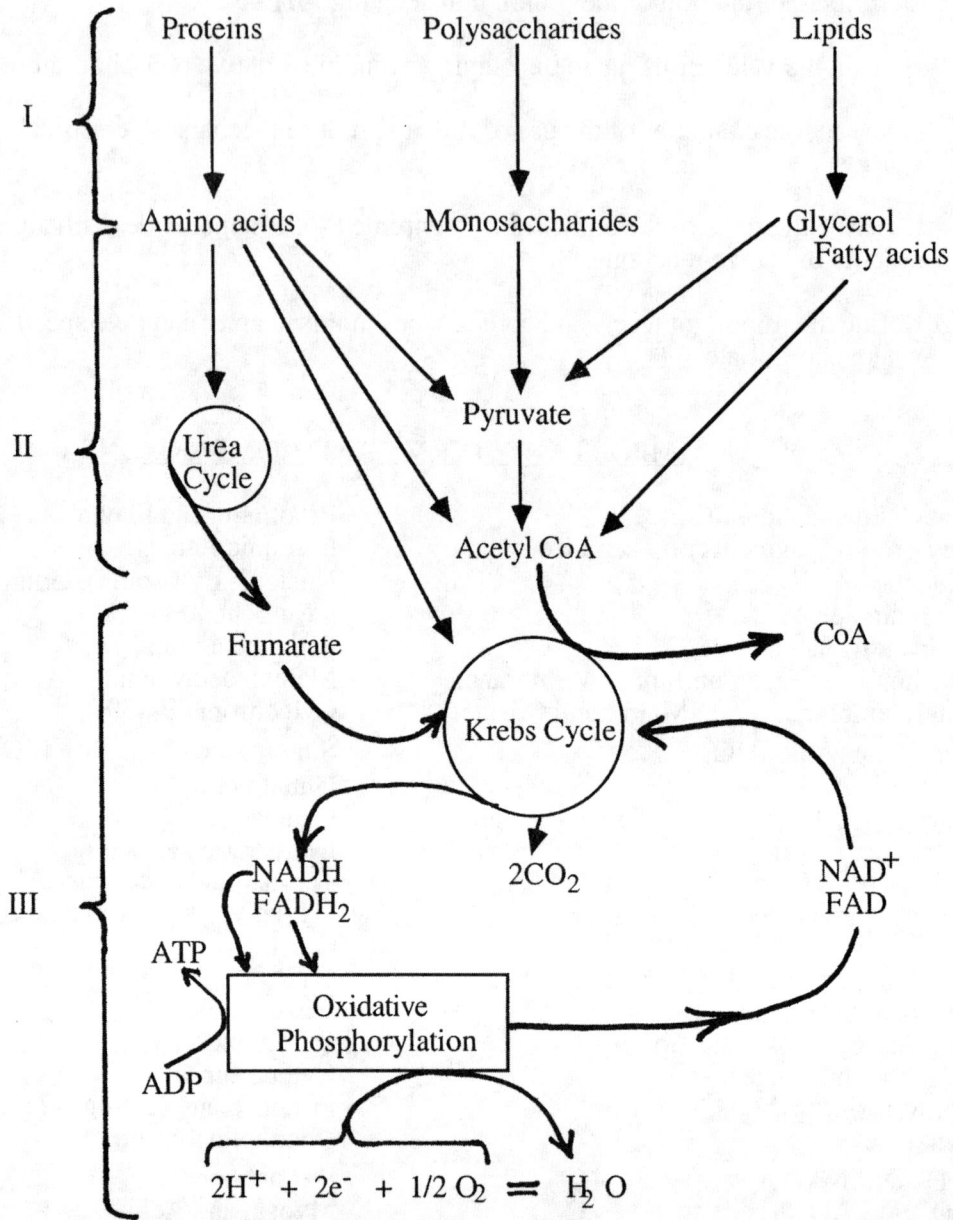

In the initial stage, the macromolecules in food are digested to monomers. Proteins are degraded to amino acids, polysaccharides to glucose and other monosaccharides and lipids are hydrolyzed to fatty acids and glycerol. In the second stage, the sugars, fatty acids, glycerol and many of the amino acids are converted into the activated acetyl units (C_2 fragments) of acetyl coenzyme A. The nitrogen atoms in amino acids are processed in the urea cycle, while the deaminated carbon skeletons are processed directly or in a few steps into the Krebs cycle. The major focus in stage III involves the Krebs cycle and oxidative phosphorylation. In these final stages of the oxidation of "fuel" molecules, the C_2 fragment of acetyl CoA enters the Krebs cycle and is completely oxidized to 2 CO_2, reduced coenzymes NADH and $FADH_2$ are produced, along with one GTP molecule.

Chapter 26 Bioenergetics: How the Body Converts Food to Energy 169

The reduced coenzymes NADH and $FADH_2$ enter in the electron transport chain and are oxidized to NAD^+ and FAD, which are then available for recycling in the Krebs cycle. As a result of the oxidation, hydrogen ions are expelled from the mitochondria and upon re-entry produce ATP from the phosphorylation of ADP. The electrons and the hydrogen ions now combine with O_2 to form water. Note that most of the ATP generated from the catabolic metabolism of the foodstuffs is formed in stage III.

II. Summary of Reactions

    a.   Interconversion of ATP to ADP + Pi

$$ATP \; \underset{\longleftarrow}{\longrightarrow} \; ADP + Pi$$

    b.   Half reactions for coenzymes NAD^+ and FAD

$$NAD^+ + H^+ + 2\,e^- = NADH$$
$$FAD + 2\,H^+ + 2\,e^- = FADH_2$$

    c.   Overall reaction in Krebs cycle

$$H_3CCOOH + 2H_2O + 3NAD^+ + FAD + GDP + Pi \longrightarrow$$

$$2CO_2 + 3NADH + FADH_2 + GTP + 3H^+$$

            waste   to electron   energy
            product   transport chain

    d.   Overall reaction of oxidative phosphorylation

$$NADH + 3ADP + 1/2\,O_2 + 3Pi + H^+ = NAD^+ + 3ATP + 4H_2O$$
$$FADH_2 + 2ADP + 1/2\,O_2 + 2Pi = FAD + 2ATP + 3H_2O$$

        from   recycle   energy
        Krebs cycle   back to
                                                  Krebs cycle

    e.   Oxidation of a C_2 (acetyl) fragment

$$C_2 + 2O_2 + 12ADP + 12\,Pi = 12\,ATP + 2CO_2$$

III. Citric Acid Cycle*

Step	Reaction	Type
1.	Acetyl CoA + oxaloacetate + H$_2$O ----> citrate + CoA (C$_4$H$_2$O$_5$) (C$_6$H$_5$O$_7$)	Condensation reaction
2.	Citrate ----> cis-aconitate + H$_2$O ----> Isocitrate (C$_6$H$_5$O$_7$) (C$_6$H$_5$O$_7$) tertiary alcohol secondary alcohol	Isomerization, by way of a dehydration, followed by a hydration
3.	Isocitrate + NAD$^+$ ----> α-ketoglutarate + NADH + CO$_2$ (C$_6$H$_5$O$_7$) (C$_5$H$_4$O$_5$) alcohol ketone	Oxidative decarboxylation
4/5	α-ketoglutarate + NAD$^+$ + H$_2$O + GDP + Pi ----> succinate + CO$_2$ + NADH + H$^+$ + GTP (C$_5$H$_4$O$_5$) (C$_4$H$_4$O$_4$) ketone acid	Oxidative decarboxylation
6.	Succinate + FAD ----> Fumarate + FADH$_2$ (C$_4$H$_4$O$_4$) (C$_4$H$_2$O$_5$)	Oxidation (of succinate)
7.	Fumarate + H$_2$O ----> malate (C$_4$H$_2$O$_4$) (C$_4$H$_4$O$_5$)	Hydration
8.	Malate + NAD$^+$ ----> oxaloacetate + NADH + H$^+$ (C$_4$H$_4$O$_5$) (C$_4$H$_2$O$_5$) alcohol ketone	Oxidation

Chapter 26 Bioenergetics: How the Body Converts Food to Energy

*The Krebs cycle can be regarded as a catalytic cycle in which a 2 carbon fragment enters the cycle in the form of acetic acid (an acetyl group plus the OH from H_2O), is oxidized and leaves the cycle as CO_2.

The major species in each step of the cycle can be easily remembered by using the following saying as a mnemonic device. "**O**f **c**ourse, **I** **k**now **s**ome **f**amous **m**en". The first letter in each substrate - **o**xaloacetate, **c**itrate, **i**socitrate, α-**k**etoglutarate, **s**uccinate, **f**umarate and **m**alate begins with these same letters as in the saying.

Again, the overall reaction in the Krebs cycle is:

$CH_3COOH + 2H_2O + 3NAD^+ + FAD + GDP + Pi \longrightarrow$

$2CO_2 + 3NADH + FADH_2\ 3H^+ +$ GTP

IV. Mitochondrion Components

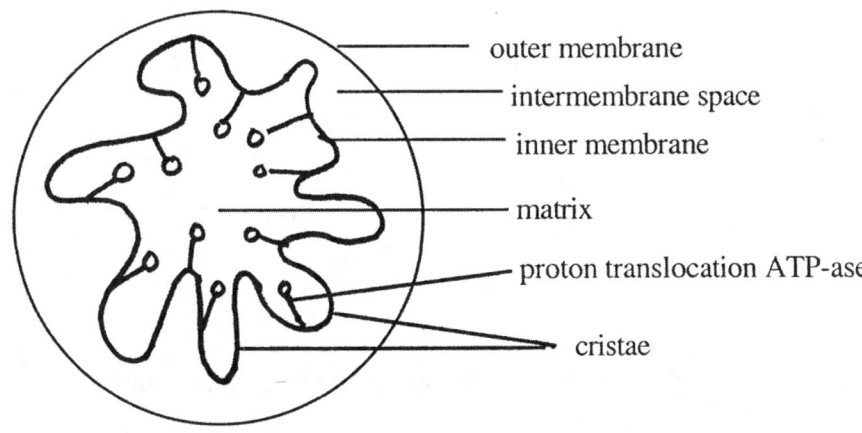

SELF-TEST QUESTIONS

<u>MULTIPLE CHOICE</u>. In the following exercises, select the correct answer from the choices listed. In some cases, two or more answers will be correct.

1. $FADH_2$ transfers its electrons into the electron transport chain in the mitochondrion directly to

    a. ATP-ase
    b. Flavoprotein
    c. Q Enzyme
    d. Cytochrome c

2. Indicate which of the following processes does not consume ATP.

    a. Muscle contraction
    b. Flow of electrons into mitochondria through proton translocating ATP-ase
    c. Catabolism
    d. Pumping of K^+ into the cells by transport proteins

Chapter 26 Bioenergetics: How the Body Converts Food to Energy

3. In the reaction in which succinate is converted to fumarate in the TCA cycle, succinate undergoes a

   a. Hydration
   b. Oxidation
   c. Isomerization
   d. Oxidative decarboxylation

4. Which of the following molecules are produced in the Krebs cycle?

   a. NAD+
   b. CO_2
   c. GDP
   d. Oxaloacetate

5. The proton translocating ATP-ase is located where in the mitochondrion?

   a. Intermembrane space
   b. Outer membrane
   c. Inside the mitochondrion
   d. Outside the inner membrane

6. The number of NADH molecules generated in the Krebs cycle is

   a. 1
   b. 2
   c. 3
   d. 4

7. The amount of energy produced by the hydrolysis of ATP is

   a. 7.3 kcal/mole
   b. 3-4 kcal/mole
   c. 32 kcal/mole
   d. less than 1 kcal/mole

8. The reduction of FAD to $FADH_2$ occurs in which part of the FAD molecule?

   a. Ribose
   b. Flavin
   c. ADP
   d. Nicotinamide

9. Indicate which of the following molecules are carriers of intermediates in metabolism.

   a. CoA
   b. ADP
   c. FAD
   d. Nicotinic acid

10. Of the following molecules, which contain high energy bonds?

    a. ATP
    b. AMP
    c. CO_2
    d. Isocitrate

11. The number of electrons, originating from the electron transport chain, that is needed to produce one mole of water is:

    a. 1
    b. 2
    c. 3
    d. 4

12. Niacin is another name for

    a. Nicotinic acid
    b. α-ketoglutarate
    c. Q enzyme
    d. Nicotinamide

Chapter 26 Bioenergetics: How the Body Converts Food to Energy   173

13. Which of the following terms is NOT another name for the Krebs cycle?

   a. Tricarboxylic acid cycle
   b. Electron transport chain
   c. Oxidative decarboxylation
   d. Citric acid cycle

14. The acetyl group in acetyl CoA is linked through what kind of a bond?

   a. (C-C) bond   c. (C-O) bond
   b. (C-S) bond   d. (C-N) bond

15. Which of the following molecules are NOT an electron-carrying enzyme in the mitochondrial membrane?

   a. Cytochrome a_3 c. Fumarase
   b. Q enzyme   d. Citrate synthetase

16. For every two C_2 fragments that enter the TCA (tricarboxylic acid) cycle, the number of ATP and (GTP) molecules produced is

   a. 24   c. 10
   b. 22   d. 12

COMPLETION. Write the word, phrase or number in the blank or draw the appropriate structure in answering the question.

1. A _____ is a compound that permits H$^+$ ions to pass through the _____ membrane passively.

2. The correct order of the enzymes, Q enzymes, FeS protein, cytochromes and flavoprotein in the electron transport chain is

   _____ , _____ , _____ , _____

3. Anabolism is associated with the _____ of molecules while catabolism deals with the _____ of molecules.

4. _____ is the molecular unit which is common in FAD, NAD$^+$, ATP and coenzyme A.

5. For each NADH molecule, a total of _____ pairs of protons are pumped out of the mitochondrion into the _____ _____ , with the resultant production of _____ molecules of ATP.

Chapter 26 Bioenergetics: How the Body Converts Food to Energy

6. Identify the 6 most important features of the mitochondrion

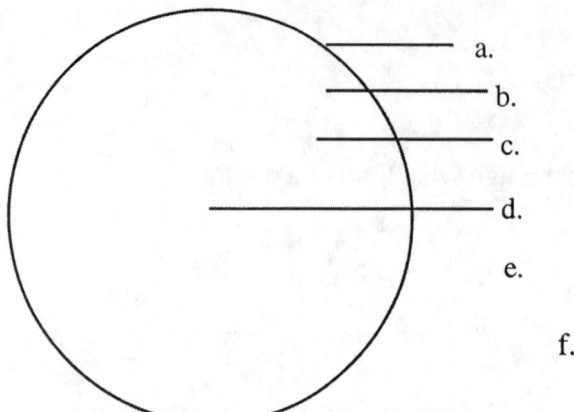

a.
b.
c.
d.
e.
f.

7. Considering the four intermediate carriers, _____ is a molecular unit unique to FAD, while _____ is a unique unit occurring in coenzyme A.

8. In the citric acid cycle, the tertiary alcohol in _____ must be converted to a secondary alcohol in _____ before the subsequent oxidation to produce a _____ group can occur.

9. The _____ hypothesis, proposed by Peter Mitchell states (in a short paragraph).

10. The reactions of the Krebs cycle occur in the _____ .

11. The products of the reaction of acetyl CoA and oxaloacetate are _____ and _____ .

12. The reactions in the electron transport chain eventually lead to the reaction of O_2 with _____ and _____ .

13. NAD^+ is a coenzyme that can act as an _____ agent in the Krebs cycle. In the process of being converted to the NADH form, it becomes _____ and picks up _____ proton and _____ electrons.

14. The major organelles and a primary function of each is:
The _____ is where DNA replication takes place.
The removal of damaged cellular components occurs in the _____.
_____ _____ are responsible for the packaging and processing of proteins which are to be secreted from the cell.

15. The two critical intracellular enzymes that are responsible for decomposing the highly reactive superoxide anion are _____ _____ and _____ .

Chapter 26 Bioenergetics: How the Body Converts Food to Energy

ANSWERS TO SELF-TEST QUESTIONS

Multiple Choice

1. c
2. b, c
3. b
4. b, d
5. c
6. c
7. a
8. b
9. a, c
10. a
11. b
12. a
13. b, c
14. b
15. c, d
16. a

Completion

1. protonphore, mitochondrial
2. flavoprotein, FeS protein, Q enzyme, cytochrome
3. synthesis, breaking down
4. ADP
5. three, intermembrane space, three
6. Refer to the figure in Section IV of Focused Review
    a. outer membrane
    b. intermembrane space
    c. inner membrane
    d. matrix
    e. proton translocating ATP-ase
    f. cristae
7. flavin, pantothenic acid
8. citric acid, isocitrate, ketone
9. chemiosmotic, the energy in the electron transfer chain creates a proton gradient. There is a higher concentration of H^+ in the intermembrane space than inside the mitochondrial matrix. The protons pumped out of the matrix provide the driving force for the phosphorylation of ADP by driving the protons back into the mitochondrial matrix through the proton translocating ATP-ase. This enzyme manufactures the ATP.
10. mitochondrion
11. citrate, CoA
12. H^+, electrons
13. oxidizing, reduced, one, two
14. nucleus, lysosomes, Golgi bodies
15. superoxide dismutase (SOD), catalase

"You are what you eat."
    - Unknown

27

Specific Catabolic Pathways: Carbohydrate, Lipid, and Protein Metabolism

CHAPTER OBJECTIVES

After you have studied the chapter and worked the assigned exercises in the text and the study guide, you should be able to do the following.

1. Name the major pathways by which energy is extracted from monosaccharides and fatty acids.

2. Discuss the three stages of glycolysis with regard to the energetics (ATP consumption or generation), the number of carbons in the molecular units and the reaction types involved.

3. Distinguish between anaerobic and aerobic glycolysis.

4. Indicate under what conditions pyruvate is converted to (a) lactate, (b) ethanol or acetyl CoA. Write out the reaction for each process.

5. Tabulate the energy yield produced by reactions involved in the complete metabolism of glucose. Indicate the influence of the glycerol-3-phosphate or the aspartate-malate shuttle in the total energy yield.

6. Indicate the point in the catabolic process at which (a) glycerol and (b) fatty acids converge to a common pathway with monosaccharides.

7. Indicate the types of reactions in the β-oxidation spiral of fatty acids.

8. Specify the cellular location in which (a) glycolysis, (b) fatty acid activation and oxidation, (c) citric acid cycle, (d) oxidative phosphorylation and the (e) urea acid cycle occur.

9. Indicate the role of the pentose phosphate pathway in general anabolic metabolism and nucleic acid metabolism, in addition to its indirect role in maintaining red blood cells.

Chapter 27 Biosynthetic Pathways

10. Outline the role of phosphorylation in the regulation of fatty acid metabolism.

11. Compare the energy yields for the complete oxidation of (a) glucose, (b) glycerol and (c) a fatty acid.

12. Name the ketone bodies and explain why they are found in high concentrations in the blood under certain conditions.

13. Discuss the three stages of nitrogen metabolism.

14. Indicate the relationship between heme, biliverdin, bilirubin, and urorubin.

15. Define the important terms and comparisons in this chapter and give specific examples where appropriate.

IMPORTANT TERMS AND COMPARISONS

Food
Glycolysis
β-Oxidation
Amino Acid Pool

Oxidative Deamination
Urea Cycle
Fat Deposit
Glucose 6-Phosphate
C_2 Fragment
Anaerobic Pathway
Pyruvate
Glycogenolysis
Biliverdin
Glycerol 3-Phosphate Transport
Aspartate-Malate Shuttle
Glycerol 1-Phosphate
Pentose Phosphate Pathway
Acyltransferase
Ubiquitin, Proteasome and Protein
  Degradation

Thiolase
Energy Yield
Ketone Bodies
Urea

α-Ketoglutarate
Carbamoyl Phosphate
Ornithine
Glucogenic and Ketogenic
  Amino Acids
Monosodium Glutamate (MSG)
Phenylketonuria (PKU)
Cytosol
Bilirubin
Ferritin
Urobilin
Jaundice
Acyl-CoA
Carnitine
Transamination

Signal Transduction and Metabolic Regulation by Phosphorylation

SELF-TEST QUESTIONS

MULTIPLE CHOICE. In the following exercises, select the correct answer from the choices listed. In some cases, two or more answers will be correct.

1. The initial reaction in the β-oxidation spiral of fatty acids produces
   a. acyl CoA
   b. acetyl CoA
   c. carbamoyl phosphate
   d. dihydroxyacetone phosphate

2. The final product of aerobic glycolysis is
   a. pyruvate
   b. acetyl CoA
   c. lactate
   d. ethanol

Chapter 27 Biosynthetic Pathways   178

3. The number of ATP molecules produced on the complete metabolism of glucose in skeletal muscle is
   a. 24
   b. 36
   c. 12
   d. 38

4. Anaerobic glycolysis occurs in the
   a. mitochondria
   b. cytosol
   c. cytosol and mitochondria
   d. extracellular fluid

5. Ketone bodies are produced in conditions of a
   a. high glucose supply
   b. low amino acid pool
   c. low glucose supply
   d. missing fructose-1-phosphate aldolase

6. The heme unit of hemoglobin contains
   a. Fe^{2+}
   b. globin protein
   c. carbohydrate
   d. glycerol

7. The first reaction involved in nitrogen catabolism of amino acids is a(n)
   a. oxidative deamination
   b. reduction
   c. hydration
   d. transamination

8. The final step in the urea cycle involves the
   a. splitting of argininosuccinate
   b. oxidation of glutamate
   c. hydrolysis of arginine
   d. transamination of glutamate

9. The order of the caloric value for carbohydrates, proteins and fats is
   a. protein = fat = carbohydrate
   b. carbohydrate > fat > protein
   c. fat > carbohydrate > protein
   d. protein > fat > carbohydrate

10. Pyruvate undergoes an oxidative decarboxylation with CoA to produce acetyl CoA and
    a. lactate
    b. glutamate
    c. CO_2
    d. NAD^+

11. Which of the following molecules is transported into the mitochondria by either the glycerol-3-phosphate transport or the aspartate-malate shuttle?
    a. ATP
    b. CO_2
    c. NADH
    d. ADP

12. The number of cycles of the β-oxidation spiral that the (saturated) palmitic acid (C_{16}) goes through is
    a. 8
    b. 16
    c. 7
    d. 4

13 A main function of the pentose phosphate pathway is
    a. generation of ribose
    b. generation of NADPH
    c. generation of NADH
    d. generation of $NADP^+$

Chapter 27 Biosynthetic Pathways   179

14. Carnitine acyltransferase is essential in the functioning of the
    a. pentose phosphate pathway   c. urea cycle
    b. fatty acid metabolism   d. ketone body production

COMPLETION. Write the word, phrase or number in the blank space in answering the question.

1. The three end products of complete catabolism of proteins in the body are _____, _____, and _____.

2. Specialized cells that store fats are called _____.

3. In the catabolic metabolism of hemoglobin, the iron is taken up by _____.

4. _____ is considered a ketone body despite not having a ketone functional group.

5. An aldolase enzyme catalyzes the splitting of fructose 1,6-diphosphate to _____ and _____.

6. Fats are metabolized by first hydrolyzing them to _____ and _____.

7. Although _____ is a basic amino acid and occurs in the urea cycle, it is not found in proteins.

8. The β-oxidation of a C_{10} unsaturated fatty acid requires one _____ step in the process than that for the saturated C_{10} fatty acid.

9. Fatty acid oxidation occurs in the _____.

10. If the carbon skeleton of an amino acid is converted to pyruvate, it may serve in either of two roles.

    a. _____

    b. _____

11. Damaged proteins are covalently modified by reaction with another protein called _____ , after which this "flagged" or modified protein is delivered to the machinery for proteolysis, called the _____.

ANSWERS TO SELF-TEST QUESTIONS

Multiple Choice
1. a
2. a
3. b
4. b
5. c
6. a
7. d
8. c
9. c
10. c
11. c
12. c
13. a, b
14. b

Chapter 27 Biosynthetic Pathways

Completion
1. urea or NH_4^+, CO_2, H_2O
2. fat deposits
3. ferritin
4. β-hydroxybutyric acid
5. glyceraldehyde 3-phosphate, dihydroxyacetone phosphate
6. glycerol, fatty acid
7. ornithine
8. more
9. mitochondrial matrix
10. a. an energy source
    b. a building block in the synthesis of glucose (and other metabolites)
11. ubiquitin, proteasome

Chapter 28   Biosynthetic Pathways

If we could not store so vital a
substance as glucose in our bodies,
we would have to be eating incessantly
to maintain a steady supply of it. That
would be a precarious existence.
    -Ernest Borek
    The Atoms Within Us

28

Biosynthetic Pathways

CHAPTER OBJECTIVES

After you have studied the chapter and worked the assigned exercises in the text and the study guide, you should be able to do the following.

1. Contrast anabolic and catabolic reactions in terms of energy requirements, the form of coenzymes involved and the location in which they generally take place.

2. Explain why anabolic reaction pathways are usually not just the reverse of the corresponding catabolic pathways.

3. Using Figure 28.1 (text), identify the major metabolic intermediates which can be used in gluconeogenesis and indicate how glycolysis differs from gluconeogenesis.

4. Discuss the importance of UDP-glucose in carbohydrate biosynthesis.

5. Identify the feature that is common in both fatty acid synthesis and degradation.

6. Describe the role of acetyl CoA and malonyl CoA in fatty acid synthesis.

7. Indicate the form of CoA and the other reactants involved in the synthesis of phosphatidate and sphingosine.

8. Write out the reactions for the synthesis of glutamic acid from α-ketoglutaric acid and the subsequent transamination reaction to form other non-essential amino acid.

9. Distinguish the difference between essential and nonessential amino acids.

Chapter 28 Biosynthetic Pathways

10. Outline the photosynthetic process in terms of a light reaction and a dark reaction.

10. Define the important terms and comparisons in this chapter and give specific examples where appropriate.

IMPORTANT TERMS AND COMPARISONS

Le Chatelier's Principle
Anabolic Pathway
Glycogen
Photosynthesis
Gluconeogenesis and Glycogenesis
Glucogenic Amino Acid
Calvin Cycle
Acyl Carrier Protein (ACP)
Essential and Non-Essential Amino Acids

Uridine Diphosphate (UDP)
Essential Fatty Acid
Synthase
Malonyl ACP
Acetyl CoA and Acyl CoA
Hyaline Membrane Disease
Light and Dark Reactions in Photosynthesis
Ri**bu**lose 1-5-**Bis**phosphate Carboxylase
-**O**xygenase (**RuBisCO**)

SELF-TEST QUESTIONS

MULTIPLE CHOICE. In the following exercises, select the correct answer from the choices listed. In some cases, two or more choices will be correct.

1. Gluconeogenesis is associated with the synthesis of
   a. glycogen
   b. glucogenic amino acids
   c. glucose
   d. glucose 1-phosphate

2. Glycolysis and gluconeogenesis proceed in reverse directions. However, the number of points in this pathway in which there are enzymes not common to each pathway is
   a. 6
   b. 8
   c. 4
   d. 2

3. Both fatty acid synthesis and degradation have the common feature of involving
   a. ATP
   b. NAD$^+$
   c. acetyl CoA
   d. ketogenic amino acids

4. In each cycle in fatty acid synthesis, the unit added on is a
   a. C_1 fragment
   b. C_2 fragment
   c. C_4 fragment
   d. C_6 fragment

5. Which of the following structures represents UDP-glucose?

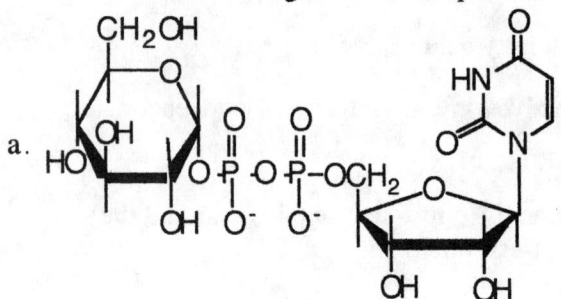

Chapter 28 Biosynthetic Pathways   183

b. 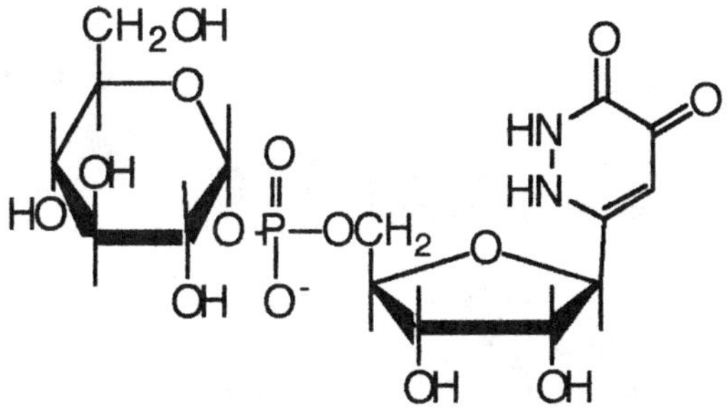 [structure]   d. 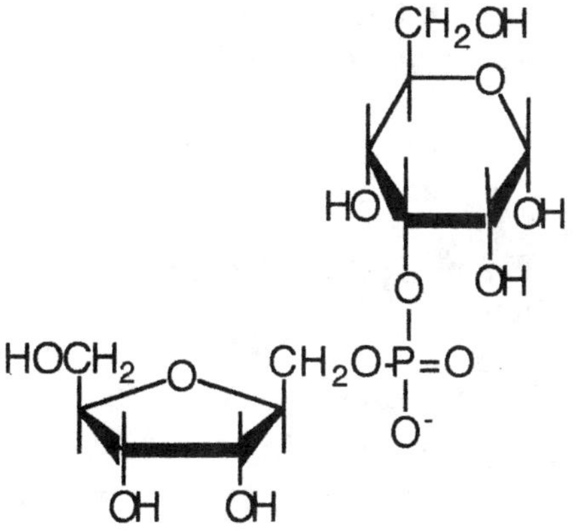 [structure]

6. The amino acid produced in the following reaction is:

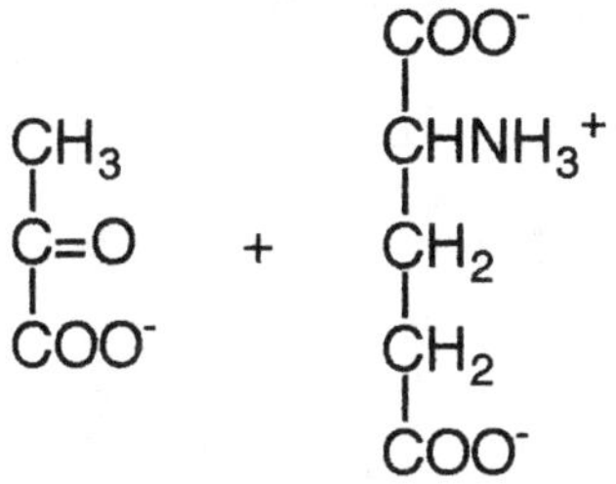

   [reaction structure: CH₃-C(=O)-COO⁻ + ⁻OOC-CH(NH₃⁺)-CH₂-CH₂-COO⁻]

   a. glycine   c. arginine
   b. glutamic acid   d. alanine

7. Which of the following can be used in gluconeogenesis?
   a. pyruvate   c. ketogenic amino acids
   b. lactate   d. urea

8. Which of the following are essential fatty acids?
   a. stearic acid   c. palmitic acid
   b. butanoic acid   d. linoleic acid

9. The Calvin cycle is associated with
   a. the dark reaction   c. utilizes RuBisCO
   b. requires UDP-glucose   d. converts H2O to O2

COMPLETION. Write the word, phrase or number in the blank space in answering the question.

1. Biosynthetic pathways require an _____ input and require _____ power in the form of NADPH.

2. Glycogen is synthesized from the activated glucose monomer, _____, which is formed in the reaction of _____ and _____.

3. The overall reaction in photosynthesis is _____.

4. Fatty acid synthesis occurs in the _____ , while fatty acid degradation takes place in the _____ .

5. The acetyl or malonyl group in the ACP complex is attached through a _____ atom.

Chapter 28 Biosynthetic Pathways

6. The fatty acid component of phosphatidate is derived from the _____ which reacts with glycerol 3-phosphate.

7. The biosynthesis of cholesterol occurs in the _____ (an organ) and is assembled from _____.

8. The phosphatidylcholine in normal lungs contains predominantly _____ fatty acids.

9. Photosynthesis occurs in the _____ organelles of plants.
   The first reaction in photosynthesis requires energy (sunlight) and is called the _____ reaction. The four important species generated are _____, _____, _____ and _____.
   The second reaction uses CO_2 and three of the species generated in the light reaction, _____, _____ and _____ to produce _____ macromolecules.

10. Most anabolic reactions occur in the _____.

11. Amino acids which can be synthesized in the human body are referred to as _____ amino acids.

ANSWERS TO SELF-TEST QUESTIONS

Multiple Choice

1. c
2. c
3. c
4. b
5. a
6. d
7. a, b
8. d
9. a, c

Completion

1. energy, reducing
2. UDP-glucose, UTP, glucose 1-phosphate
3. $n\ H_2O\ +\ n\ CO_2\ \xrightarrow{\text{sunlight}}\ (CH_2O)_n\ +\ n\ O_2$
4. cytosol, mitochondria
5. S
6. acyl CoA
7. liver, acetyl CoA
8. unsaturated
9. chloroplast;
   light, ATP, NADPH, H^+, O_2;
   ATP, NADPH, H^+, carbohydrate
10. cytoplasm
11. nonessential

"I hate quotations. Tell me what you know."
 - Ralph Waldo Emerson

29

Nutrition and Digestion

CHAPTER OBJECTIVES

After you have studied the chapter and worked the assigned exercises in the text and the study guide, you should be able to do the following.

1. Define nutrient and classify the nutrients into the six groups.

2. Indicate what RDA stands for.

3. Explain the difference between basal caloric requirement and the normal caloric intake of active individuals.

4. Explain why a lengthy diet is usually necessary to lower weight by more than a few pounds.

5. Indicate the primary roles in the body for carbohydrates, fats and proteins.

6. Distinguish between digestible and non-digestible carbohydrates.

7. Distinguish between a complete and an incomplete protein from a nutritional standpoint.

8. Define generally the process of digestion.

9. Outline the digestion process for carbohydrates and point out the role of the most important enzymes involved.

10. Outline the digestion process for lipids and point out the role of the most important enzymes involved.

11. Outline the digestion process for proteins and point out the role of the most important enzymes involved.

Chapter 29   Nutrition and Digestion

12. Name the water-soluble vitamins and indicate one function for each.

13. Name the fat-soluble vitamins and indicate one function for each

14. Define the important terms and comparisons in this chapter and give specific examples where appropriate.

IMPORTANT TERMS AND COMPARISONS

Nutrients
Diabetes Mellitus
Basal Caloric Requirement
Essential Amino Acids
Kwashiorkor
Bile Salts
Fiber
Hydrolysis Reaction
Debranching Enzyme
α- and β-Amylase
Blood Sugar
Parenteral Nutrition
Fat-Soluble Vitamins
  Vitamin A
  Vitamin D
  Vitamin E
  Vitamin K
Water
Major Minerals
Trace Elements
Obesity
Artificial Sweeteners
Dietary Performance Enhancers

Nutritional Calorie
Complete Protein
Minerals
Cellulase and Cellulose
Marasmus
Lipases
Digestion
Acetal Linkage and Peptide Linkage
Hypoglycemia and Hyperglycemia
Digestive Proteolytic Enzymes
  Chymotrypsin, Pepsin
  Trypsin, Carboxypeptidase
Water-Soluble Vitamins
  Vitamin C (Ascorbic Acid)
  Vitamin B_1 (Thiamine)
  Vitamin B_2 (Riboflavin)
  Vitamin B_6 (Pyridoxal)Niacin
  Vitamin B_{12}
  Niacin
  Folic Acid
  Pantothenic Acid
  Biotin

SELF-TEST QUESTIONS

MULTIPLE CHOICE. In the following exercises, select the correct answer from the choices listed. More than one correct answer may be possible in these questions.

1. Which of the following are classified as nutrients?
   a. carbohydrates
   b. nucleic acids
   c. esters
   d. vitamins
   e. water

2. A nutrient calorie (Cal.) is equivalent to
   a. 100 calories
   b. 1 calorie
   c. 1000 calories
   d. no fixed number because it will depend on the route of digestion

3. The most prevalent form of dietary fats i
   a. fatty acids
   b. cholesterol
   c. triglycerides
   d. sphingolipids

Chapter 29   Nutrition and Digestion

4. The essential fatty acids are
    a. linolenic and linoleic acids
    b. acetic and linoleic acids
    c. palmitic and linolenic acids
    d. stearic and linoleic acids

5. Water makes up what percent of body weight
    a. 15%
    b. 75%
    c. 35%
    d. 60%

6. A number of enzymes are necessary for the digestion of carbohydrates. The role of β-amylase is to
    a. hydrolyze the α (1-4) glycosidic linkages in an orderly fashion, cutting disaccharide maltose units one by one from the nonreducing end of the chain
    b. randomly cut the α (1-4) glycosidic linkages
    c. hydrolyze the α (1-6) glycosidic linkage
    d. hydrolyze all glycosidic linkages, irrespective of the type of linkage

7. A carbohydrate that cannot be hydrolyzed in our digestive system is
    a. glycogen
    b. cellulose
    c. amylose
    d. amylopectin

8. Some of the final digestion products of lipids as a result of lipase action are
    a. glycerol
    b. choline
    c. phospholipids
    d. sphingolipids

9. Digestion of proteins results from the hydrolysis of the
    a. phosphodiester linkage
    b. peptide linkage
    c. glycosidic backbone
    d. hemiacetal linkage

10. Enzymes involved in the digestion of proteins include
    a. pepsin
    b. lipase
    c. carboxypeptidase
    d. α-amylase

11. The water soluble vitamins include
    a. vitamin B_{12}
    b. biotin
    c. thiamine
    d. niacin

12. The only fat-soluble vitamin in the following list is
    a. vitamin K
    b. niacin
    c. ascorbic acid
    d. folic acid

COMPLETION. Write the word, phrase or number in the blank space in answering the following questions.

1. In order to lose 10 pounds of fat, a person must use up _____ Cals.

2. We must consume certain amounts of vitamins, minerals and nutrients. Typically, we must take in _____ of nutrients, while we need only _____ of vitamins and minerals.

Chapter 29 Nutrition and Digestion

3. Most of the minerals are metal ions. Five of the metal ions are _____, _____, _____, _____, and _____.

4. Two minerals that are not metals are _____ and _____.

5. Two fat-soluble vitamins are _____ and _____.

6. Four water soluble vitamins are _____, _____, _____ and _____.

7. Linolenic and linoleic are essential fatty acids in higher animals. They both contain _____ carbons in the make-up of the fatty acid.

8. Three water soluble vitamins which act as coenzymes in biological reactions are _____, _____ and _____.

9. _____ is a complete protein, while examples of proteins that are not complete are _____ and _____.

10. _____ are the most concentrated form of energy that we consume as food.

11. Olestra is effectively a fat since is has _____ functional groups. It is derived from the sugar, _____, in which the ____ groups have been esterified by reaction with long chain fatty acid groups.

MATCHING. Match the term in the right-hand column with the appropriate answer in the left-column.

1. Vitamin A
2. Thiamine
3. Ascorbic Acid
4. Vitamin K
5. Vitamin E
6. Vitamin D
7. Folic Acid
8. Vitamin B_{12}

a. Fat-soluble antioxidant
b. Found in leafy vegetables
c. Contains the metal, cobalt
d. Another name for vitamin B_1
e. Important in bone formation
f. Prevalent in citrus fruit
g. Important for blood clotting
h. Important for proper vision

ANSWERS TO SELF-TEST QUESTIONS

Multiple Choice
1. a, d, e
2. c
3. c
4. a
5. d
6. a
7. b
8. a, b, d
9. b
10. a, c
11. a, b, c and d
12. Vitamin K

Completion
1. 35,000 Cal.
2. (many) grams; micro- or milligrams
3. Any five of Na, Ca, Mg, Fe, Zn, Cu, Mn, Cr, Mo or Cu
4. Any two of Cl, P, Se, I or F.
5. Any two of vitamins A, D, E or K.

Chapter 29 Nutrition and Digestion

6. Any four of vitamins B_1, B_2, B_6, B_{12}, niacin, folic acid, pantothenic acid, biotin, or vitamin C.
7. 18
8. Any 3 of vitamins B_1, B_2, B_6, B_{12}, niacin or folic acid.
9. Casein; Any two of gelatin or corn, rice or wheat protein.
10. Fats
11. ester, sucrose, -OH

Matching
1. h
2. d
3. f
4. g
5. a
6. e
7. b
8. c

"I hate quotations. Tell me what you know."
- Ralph Waldo Emerson

30

Immunochemistry

CHAPTER OBJECTIVES

After you have studied the chapter and worked the assigned exercises in the text and the study guide, you should be able to do the following.

1. Indicate the characteristics that distinguish innate immunity and acquired immunity.

2. Outline the role of B cells in the development of an immune response.

3. Outline the role of the T cells in the development of an immune response.

4. Indicate the characteristics of killer and memory T cells.

5. Describe the interaction between an antigen and an antibody, emphasizing the molecular interactions involved on each.

6. Describe the function of Class I and Class II major histocompatibility complexes (MHC) in the immune response.

7. Distinguish between the different classes of immunoglobulins in terms of size, function and location.

8. Characterize the variable and constant regions of an immunoglobulins in terms of function and the region within the protein.

9. Characterize the different types of cytokines

10. Define the important terms and comparisons in this chapter and give specific examples where appropriate.

Chapter 30   Immunochemistry

IMPORTANT TERMS AND COMPARISONS

Innate Immunity - External and Internal
Lymphocytes - B Cells and T Cells
Antigens and Antibodies
Immunoglobulin Super Family
Macrophages
NO Synthase
Killer and Memory T Cells
Antibodies and Immunogens
Major Histocompatibility Complex (MHC)
IgG, IgE, IgA, IgD and IgE
Antigen-Antibody Complex
Cluster Determinant (CD) Proteins
Glycoprotein 120 (gp120)
T Cell Receptor Complex

Acquired or Adaptive Immunity
Antibodies and Immunoglobulins
Signaling Proteins - Cytokines
Antibodies, T-Cell Receptors, Surface
  Presentation of Antigen Peptide
T Cell Differentiation
Plasma Cells
Epitope
Classes of Immunoglobulins
Large and Small Chain (Subunits)
Monoclonal Antibodies
T-Cell Receptor (TcR)
Interleukins and Chemokines

SELF-TEST QUESTIONS

MULTIPLE CHOICE. In the following exercises, select the correct answer from the choices listed. More than one correct answer may be possible in these questions.

1. The initial binding of a foreign molecule is carried out by
   a. macrophages
   b. CD4
   c. MHC
   d. B cells

2. An epitope is found on
   a. an immunoglobulin protein
   b. a T-cell surface protein
   c. an antigen
   d. a cytokine

3. The role of the MHC is to
   a. bind CD4
   b. stimulate cytokine formation
   c. bind to antigen in plasma
   d. bringing the antigen's epitope to the cell surface

4. The smallest (by MW) antibody in the blood is
   a. IgG
   b. IgD
   c. IgE
   d. IgM

5. The number of polypeptide chains in an immunoglobulin is
   a. one
   b. four
   c. two
   d. six

6. The region of the polypeptide chains in antibodies that interacts directly with the antigen is
   a. N-terminal end
   b. C-terminal end
   c. central region only
   d. distributed over many regions

7. B cells are stimulated to differentiate into plasma cells by
   a. exposure to cytokines
   b. antigen binding to the surface-bound antibody
   c. on interaction with macrophages
   d. after MHC display on surface

8. Each B cell is capable of synthesizing
   a. many different antibodies
   b. an unknown number that depends on the situation
   c. only one immunoglobulin
   d. more than one, but clearly a limited number

9. T cells bind to antigen by the surface
   a. TcR complex
   b. TcR and cytokine together
   c. TcR
   d. CD4 receptor

10. Monoclonal antibodies are derived from
    a. a single B cell
    b. multiple B cells
    c. a single T cell
    d. a specific type of macrophage

11. Cytokines are proteins that are modified by
    a. sugar groups (glycoproteins)
    b. acetyl groups
    c. phosphate groups
    d. methyl groups

COMPLETION. Write the word, phrase or number in the blank space in answering the following questions.

1. Myasthenia gravis is an _____ disease in which the patient develops antibodies against _____ receptors.

2. Heceptin is a _____ antibody that is used in the treatment of breast cancer and binds to the antigen called _____.

3. A immunization shot produces only a _____ and is not effective. A second shot, the "booster shot", establishes the immunization by increasing its effective because the _____ cells divide and become both _____ and _____ cells.

4. A key difference between innate and acquired immunity is that the former exhibits no _____ or _____, while these characteristics are the hallmark of acquired immunity.

5. Of the two major types of T cell, only the _____ T cells release the protein, _____, which effectively plugs a hole in the target cell.

ANSWERS TO SELF-TEST QUESTIONS
Multiple Choice
1. a
2. c
3. d
4. a
5. b
6. a
7. b
8. c
9. a
10. a
11. a

Completion
1. autoimmune, acetylcholine
2. monoclonal
3. short term response, memory, memory, plasma
4. specificity, memory
5. killer, perforin

Chapter 30 Immunochemistry

IMPORTANT TERMS AND COMPARISONS

Innate Immunity - External and Internal
Lymphocytes - B Cells and T Cells
Antigens and Antibodies
Immunoglobulin Super Family
Macrophages
NO Synthase
Killer and Memory T Cells
Antibodies and Immunogens
Major Histocompatibility Complex (MHC)
IgG, IgE, IgA, IgD and IgE
Antigen-Antibody Complex
Cluster Determinant (CD) Proteins
Glycoprotein 120 (gp120)
T Cell Receptor Complex

Acquired or Adaptive Immunity
Antibodies and Immunoglobulins
Signaling Proteins - Cytokines
Antibodies, T-Cell Receptors, Surface
    Presentation of Antigen Peptide
T Cell Differentiation
Plasma Cells
Epitope
Classes of Immunoglobulins
Large and Small Chain (Subunits)
Monoclonal Antibodies
T-Cell Receptor (TcR)
Interleukins and Chemokines

SELF-TEST QUESTIONS

MULTIPLE CHOICE. In the following exercises, select the correct answer from the choices listed. More than one correct answer may be possible in these questions.

1. The initial binding of a foreign molecule is carried out by
   a. macrophages
   b. CD4
   c. MHC
   d. B cells

2. An epitope is found on
   a. an immunoglobulin protein
   b. a T-cell surface protein
   c. an antigen
   d. a cytokine

3. The role of the MHC is to
   a. bind CD4
   b. stimulate cytokine formation
   c. bind to antigen in plasma
   d. bringing the antigen's epitope to the cell surface

4. The smallest (by MW) antibody in the blood is
   a. IgG
   b. IgD
   c. IgE
   d. IgM

5. The number of polypeptide chains in an immunoglobulin is
   a. one
   b. four
   c. two
   d. six

6. The region of the polypeptide chains in antibodies that interacts directly with the antigen is
   a. N-terminal end
   b. C-terminal end
   c. central region only
   d. distributed over many regions

7. B cells are stimulated to differentiate into plasma cells by
   a. exposure to cytokines
   b. antigen binding to the surface-bound antibody
   c. on interaction with macrophages
   d. after MHC display on surface

Chapter 30   Immunochemistry

8. Each B cell is capable of synthesizing
   a. many different antibodies
   b. an unknown number that depends on the situation
   c. only one immunoglobulin
   d. more than one, but clearly a limited number

9. T cells bind to antigen by the surface
   a. TcR complex
   b. TcR and cytokine together
   c. TcR
   d. CD4 receptor

10. Monoclonal antibodies are derived from
    a. a single B cell
    b. multiple B cells
    c. a single T cell
    d. a specific type of macrophage

11. Cytokines are proteins that are modified by
    a. sugar groups (glycoproteins)
    b. acetyl groups
    c. phosphate groups
    d. methyl groups

COMPLETION. Write the word, phrase or number in the blank space in answering the following questions.

1. Myasthenia gravis is an _____ disease in which the patient develops antibodies against _____ receptors.

2. Heceptin is a _____ antibody that is used in the treatment of breast cancer and binds to the antigen called _____.

3. A immunization shot produces only a _____ and is not effective. A second shot, the "booster shot", establishes the immunization by increasing its effective because the _____ cells divide and become both _____ and _____ cells.

4. A key difference between innate and acquired immunity is that the former exhibits no _____ or _____, while these characteristics are the hallmark of acquired immunity.

5. Of the two major types of T cell, only the _____ T cells release the protein, _____, which effectively plugs a hole in the target cell.

ANSWERS TO SELF-TEST QUESTIONS
Multiple Choice
1. a
2. c
3. d
4. a
5. b
6. a
7. b
8. c
9. a
10. a
11. a

Completion
1. autoimmune, acetylcholine
2. monoclonal
3. short term response, memory, memory, plasma
4. specificity, memory
5. killer, perforin

"The thing to remember when traveling
is that the trail is the thing, not the end
of the trail. Travel too fast and you miss
all you are traveling for."
- Louis L'Amour

31

Body Fluids

CHAPTER OBJECTIVES

After you have studied the chapter and worked the assigned exercises in the text and the study guide, you should be able to do the following.

1. Tabulate the functions of blood and indicate which molecules play a major role in these functions.

2. Characterize each of the three main blood cell types, including where they are formed and their major functions.

3. Compare the components of blood, blood plasma, and serum.

4. Discuss the interrelationship between prothrombin, thrombin, thromboplastin, and thrombosis.

5. Outline the characteristics of the hemoglobin molecule, its O_2 binding characteristics and how the CO_2 and H^+ influence O_2 binding.

6. Explain the role of nephrons in retaining proteins and vital nutrients, while eliminating waste products from the blood.

7. Discuss the role of the hormones vasopressin and aldosterone in the control of water and salt balance in the blood and kidneys.

8. Draw the general structure of an antibody and indicate the variable regions which interact with antigens.

9. Define the important terms and comparisons in this chapter and give specific examples where appropriate.

Chapter 31   Body Fluids

IMPORTANT TERMS AND COMPARISONS

Extracellular, Interstitial and Intercellular Fluid
Erythrocytes, Leukocytes and Platelets
Plasma and Serum
Blood Clotting
Fibrinogen and Fibrin
Prothrombin and Thrombin
Thromboplastin
Hemoglobin
Oxygen Dissociation Diagram
Cooperative Action
Synthetic Blood Substitute
Carbaminohemoglobin
Baroreceptor
Carbonic Anhydrase

LeChatelier's Principle
Acidosis and Alkalosis
Blood-Brain Barrier
Bowman's Capsule
  Tubule
Proximal Tubule, Henle's Loop and Distal
  Tubule
Urine, Diuresis and Antidiuretic
Vasopressin
Arterial and Venous Capillaries
Renin, Angiotensinogen and Angiotensin
Homeostasis
Bohr Effect
Tachycardia

SELF-TEST QUESTIONS

MULTIPLE CHOICE. In the following exercises, select the correct answer from the choices listed. In some cases, two or more choices will be correct.

1. The plasma contains
   a. albumin
   b. erythrocytes
   c. hemoglobin
   d. globulin

2. The synthetic blood substitute containing perfluorocarbon
   a. dissolves O_2
   b. dissolves CO_2
   c. contains antigens
   d. contains clotting factors

3. The Bohr effect is associated with the
   a. blood filtration in the kidneys
   b. action of hormones in the blood
   c. salt balance in the cell
   d. pH effect on O_2 binding to hemoglobin

4. Blood clotting can be prevented during transfusions by the addition of sodium citrate. This agent
   a. complexes with Ca^{2+}
   b. converts thrombin back to prothrombin
   c. produces edema
   d. degrades collagen

5. Extracellular fluids are those
   a. within the cell
   b. not inside the cell
   c. found in cytosol
   d. with little or no Cl^-

6. A molecule of CO_2 binds to hemoglobin to form carbaminohemoglobin. The CO_2 forms a covalent bond at the
   a. Fe(II) in the heme group
   b. amino-terminal end of the polypeptide chain
   c. carboxyl-terminal end of the polypeptide chain
   d. H^+ binding site

Chapter 31 Body Fluids

7. Carbonic anhydrase catalyzes which of the following reactions.
   a. $Hb + O_2 \rightleftarrows HbO_2$ Hb = hemoglobin
   b. $CO_2 + H_2O = H_2CO_3$
   c. $HbO_2 + H^+ + CO_2 = Hb(CO_2)H^+ + O_2$
   d. $H_2PO_4^- = H^+ + HPO_4^{2-}$

8. Vasopressin is
   a. a diuretic hormone
   b. an antidiuretic hormone
   c. a small peptide
   d. manufactured in the hypophysis

9. Hemoglobin more readily binds to the second, third and fourth O_2 molecules than it does the first because of
   a. the Bohr effect
   b. an allosteric effect
   c. a pH effect
   d. Henle's loop

10. Some of the functions of the blood include
    a. maintaining the pH of the body
    b. fighting infection
    c. carrying nutrients
    d. carrying O_2 from the lungs to the tissues

11. The cells that are made in the bone marrow include
    a. erythrocytes
    b. leukocytes
    c. platelets
    d. nephrons

12. An effective anticoagulant in the blood
    a. thins the plasma
    b. complexes with Ca^{+2} ions
    c. decreases the number of erythrocytes
    d. reduces the chloride concentration

13. The hemocrit level in a blood sample is the
    a. percent of erythrocytes in the blood
    b. oxygen (gas) pressure
    c. percent of leukocytes in the blood
    d. percent of oxygenated hemoglobin in the blood

COMPLETION. Write the word, phrase or number in the blank space in answering the question.

1. Severe dehydration stimulates the production of the hormone _____ to enable the body to retain more _____.

2. Hormonal control of blood pressure involves the secretion of the enzyme _____ from the kidney. This enzyme converts the inactive protein, _____, into the potent vasoconstrictor, _____.

3. The serum contains all the elements of plasma except for _____.

4. Persons with hypertension have _____ blood pressure. This can lead to a number of medical problems, including _____ or _____.

5. The _____ molecule is bound to the _____ in the fully

Chapter 31 Body Fluids   198

oxygenated form of hemoglobin.

6. _____ is the major component of the blood which is a liquid and contains no _____.

7. _____ is an _____ which initiates blood clotting. It is ordinarily found in the blood in the inactive form called _____.

8. Most of the CO_2, a waste product in cellular metabolism, is transported to the lungs as _____.

9. The _____ organ acts a superfiltration unit, in which the waste products are eliminated in the _____.

10. _____ is the name of the condition which occurs when the pH of the blood increases above normal.

11. A nephron contains _____ capsule which is connected to a tiny tube called the _____. The U-shaped twist in the unit is called _____ _____.

ANSWERS TO SELF-TEST QUESTIONS

Multiple Choice
1. a, d
2. a, b
3. d
4. a
5. b
6. b
7. b
8. b, c, d
9. b
10. a, b, c, d
11. a, b, c
12. b
13. a

Completion
1. aldosterone; water
2. rennin; angiotensinogen; angiotensin
3. fibrinogen
4. high; heart attack; stroke; kidney failure (any two of these)
5. O_2; Fe in the heme group
6. plasma; cells
7. thrombin; enzyme; prothrombin
8. H_2CO_3
9. kidney, urine
10. alkalosis
11. Bowman's; tubule; Henle's loop

NOTES

NOTES

NOTES